Contents

D1741209

6

 **Introduction**

This book is intended for a particular group of children. It is aimed specifically at those children who are having difficulty with their number work. Commonly, the children will not yet have achieved the competencies in Number normally expected in a ten-year-old child. These are children who require an individual education plan under the arrangements for children with special educational needs. The book outlines the skills, the teaching strategies, the activities and the follow-up work appropriate in this context.

How to use this book

There are two ways in which this book can be used in the classroom.

1 It can be used as the basis of an individual education plan. There may be several children in your class who require the provision of these plans, which can be constructed using the outline of topics and the structure of skills provided in this book. The best method here is to adopt a four-stage approach:

- identify and list the precise skills you feel the child needs to acquire
- use the Numeracy Support Book to help you construct an order of work, giving a progression from one skill to the next
- select the teaching strategies and tactics in each topic that most appropriately match the specific needs of the child
- identify and prepare the activities and follow-up work.

2 The book can be also used as a resource to help you provide focused support for particular number skills. Some children may be having difficulty with certain skills, e.g. subtraction, and need specific attention. This book supplies teaching strategies and tactics to help you to assist the children's acquisition of these skills, and the activities necessary for rehearsal and reinforcement. In this sense, it is possible to 'dip into' the book, using it as a resource when necessary, either on a 'one-off' basis or to help a small group of children.

Structure and content

The book is divided into sections under numerical skills headings. In each section, you will find:

1 Teaching skills

The skills provide a step-by-step progression for helping children to understand and assimilate a particular piece of mathematics. They are outlined briefly in a clear and logical progression which is easy for the teacher to scan and absorb. These skills inform the nature of the assistance given to the children as they attempt the activities and the follow-up work.

2 Tactics

The tactics offer a range of different practical techniques for helping children to learn. A teacher may well select those from the list that s/he finds most appropriate. Factors that will influence the choice of tactics will include the resources available in the classroom, the nature of the difficulties experienced by the child, and the particular preferences of the teacher.

3 Activities

The activities are quick and easy to prepare and straightforward to teach. Most of the activities use materials provided in the Resource Bank (e.g. number cards), so that the teacher does not have to spend time making extra materials. The children can record their efforts easily where appropriate. Each activity can be done by one child working alone, or by two or three children working together. The activities are arranged in an approximate order of difficulty.

4 Follow-up work (photocopy masters)

The follow-up work provides carefully graded examples of the specific numerical skills. The steps between one example and the next have been made as small as possible, and there is a high level of repetition. It is suggested that the children are encouraged to use the tactics outlined in each section to complete the sheets.

Further information

The Numeracy Support Book provides a detailed breakdown, in each topic, of the progression and structure of skill acquisition. Taking each numerical topic in turn, the book constructs a skill-hierarchy, outlining the steps which the child must accomplish in order to acquire that skill. Alongside this structure are the precise tactics that will assist the child with each step.

The analysis of each numerical skill can be extremely useful information for any teacher involved in planning a detailed programme of study, either for an individual, for a group or for the whole class.

1 Numbers up to 100 000

Skills

Step 1 Start with numbers up to 100. Ensure that children can place these in order, particularly where they cross a decade (e.g. 29, 30, 31), and that they can always identify the larger and the smaller number in a pair.

Step 2 Add a number of hundreds to a 2-digit number to make a 3-digit number. Emphasise that the size of the hundreds digit tells us how large a 3-digit number is.

Step 3 Compare two 3-digit numbers by looking first at the hundreds then at the tens.

Step 4 Add a number of thousands to a 3-digit number. Show how to build up a number by covering the hundreds and thousands and reading the 2-digit number, then uncovering the hundreds for the 3-digit number, and finally uncovering the thousands for the 4-digit number.

Step 5 Introduce the inequality signs (< and >) using a 'crocodile's mouth' as the image. The crocodile always faces the larger fish!

Step 6 Move on to larger numbers, adding tens of thousands if appropriate. Continue to build up the number, uncovering the tens of thousands last to read the 5-digit number. Stress the number of thousands as a 2-digit number of thousands.

Tactics

Place-value cards

These are indispensable for helping children to build up numbers. Help children to build a number starting with the two digits of lowest value. So they start with 64, reading this as: *sixty-four*. They then choose a hundreds card, e.g. 400, to make 464. They should read this aloud: *four hundred and sixty-four*. Next add the thousands, e.g. 9000. *Nine thousand four hundred and sixty-four*. When the children are confident with building up a number this way, they can include tens of thousands.

Encourage them to create 5-digit numbers with no hundreds, tens or units, and practise reading the tens of thousands as a 2-digit number of thousands, e.g. 36 000. *Thirty-six thousand*. Compare two numbers of this type to show that the 2-digit number of thousands is used for comparing. E.g. 36 000 is bigger than 35 000, because 36 is bigger than 35.

Number grid (100 to 1000)

Create a 10 × 10 number grid (100, 110, 120, ... 980, 990) where the numbers rise in tens.

Stress the multiples of 100 – help children to remember the order by chanting them. Choose another starting number from the top row of the grid and count in hundreds. E.g. *One hundred and sixty, two hundred and sixty, ...* Choose numbers on the grid and ask the children to say the number 100 more, and 100 less.

Once children are secure with 3-digit numbers, it is relatively easy to progress to 4-digit numbers. Choose a 3-digit number from the grid and write it down. Add a number of thousands, e.g. 360 → 4360.

Crocodile signs

Draw the inequality signs as the mouth of a crocodile. Draw two numbers as fish. Show the children that the crocodile always faces the largest fish.

Activities

Note: All the activities can be undertaken either by individuals or by small groups. The activities are given here in increasing order of difficulty.

ACTIVITY ❶ **Number cards (0 to 99), a number grid (0 to 99), a blank 10 × 10 grid (100 in the top left-hand corner, 990 in the bottom right-hand corner).**
Shuffle the cards and place them face down in a pile.

Take a card and find the matching number on the 0–99 number grid.

Write this number with an extra zero to turn it into a 3-digit number. Write this 3-digit number in the correct place on the blank grid.

When all the spaces in the blank grid have been completed, compare both grids: 0–99 and 100–990.

ACTIVITY ❷ **A list of 3-digit numbers**
Look at the 3-digit numbers. Write the units neighbours, the tens neighbours and the hundreds neighbours for each.

E.g. 472   its units neighbours are 473 and 471
            its tens neighbours are 482 and 462
            its hundreds neighbours are 572 and 372

ACTIVITY ❸ **Place-value cards (Th, H, T, U)**
Shuffle each set of cards and place them face down in four piles.

Take a card from three of the piles and make a number with a zero in one of its places, e.g. 4026. Continue until there are twelve numbers laid out on the table.

Write the numbers in order, smallest to largest.

ACTIVITY ❹ **Four dice**
Throw all four dice. Make the largest 4-digit number possible and write it down. Make the smallest 4-digit number possible and write it down.

Repeat five times. What is the largest number you have written? What is the smallest?

ACTIVITY ❺ **Two sets of place-value cards (Th, H, T, U), sign cards (<, >, =, +)**
Use the place-value cards to make two 4-digit numbers. Compare them using the sign cards, e.g. 3971 > 3471.

Once you are confident doing this, make some additions using the 1000 card, and two hundreds cards, e.g. 1000 = 400 + 600.

Photocopy Masters
Numeracy Support Book
pages 24 to 27

Associated units
Abacus 6 N1 Place-value

2 Adding and subtracting tens and hundreds

Skills

Children who try to complete additions such as 46 + 29 using fingers, or by trying to perform a written algorithm in their heads are missing some crucial strategies for mental fluency.

Step 1 Rehearse counting in tens. Children can use either verbal or visual clues to help them remember the chant. They may 'hear' the rhyme or 'see' the pattern: 24, 34, 44, 54, ...

Step 2 Add 10 or a multiple of 10 to any 2-digit number. Present numbers at random and ask the child to add 10, e.g. 43 + 10, 79 + 10. Progress to multiples of 10, e.g. 43 + 30, 79 + 20.

Step 3 Add numbers with a unit digit of 1 (e.g. 21, 31) to 2-digit numbers. First add the multiple of 10, then count on 1, e.g. 43 + 21 = 43 + 20 + 1.

Step 4 Add numbers with a unit digit of 9 (e.g. 29, 39) to 2-digit numbers. First add the next multiple of 10, then count back 1, e.g. 43 + 29 = 43 + 30 – 1.

Step 5 Add 100 and multiples of 100 to 3-digit numbers using the same steps (first 347 + 200, then 347 + 201, then 347 + 199).

Tactics

Number grids (0 to 99 and 100–990)

It is essential that children have an image in their minds of the number grid 0 to 99, and also of the extension grid 100 to 990. Present them with either grid as often as possible and ask them to colour or copy the series of numbers in any one column, e.g. 6, 16, 26, 36, ..., or 450, 550, 650, 750.... Encourage them to see the patterns. *Counting in tens, only the tens digits change. Counting in hundreds, only the hundreds digits change.*

Start with the 0 to 99 grid so that the children gain in confidence. Once they have mastered counting in tens and adding multiples of 10, relate this to counting in hundreds and adding multiples of 100.

Cassette

Many children find the verbal pattern of sequences of numbers such as 24, 34, 54, 64, ... and 254, 354, 454, ... easier to remember than the visual pattern.

Use a cassette and ask the children to record different number sequences of adding 10 and adding 100. They can play these back and check each sequence against the appropriate number grid.

Calculator

Reinforce the patterns of adding 10 and 100 using a calculator. Utilise the constant function so that the calculator adds 10 or 100 each time the '+' or the '=' button is pressed. The children can write down each sequence. Since the hard work of the calculation is removed, children are far more likely to see the patterns. Focus particularly on adding tens to 3-digit numbers.

Activities

Note: All the activities can be undertaken either by individuals or by small groups. The activities are given here in increasing order of difficulty.

ACTIVITY ❶

Place-value cards (hundreds, tens and units), a dice

Shuffle the cards separately and place them face down in two piles.

Take a card from each pile and make a 2-digit number (e.g. 45). Throw a dice (e.g. 3) and add that number of tens (i.e. 30). Record the addition as 45 + 30 = 75.

Repeat until all the cards are used.

ACTIVITY ❷

Number grid (100 to 990), counters, a dice

Choose a number, cover it with a counter and write it down, e.g. 320. Throw the dice (e.g. 6) and write that number of hundreds (i.e. 600). Add the two numbers and complete the addition: 320 + 600 = 920.

Repeat ten times, trying to place a counter in each row and column.

Which addition gives an answer closest to 1000?

ACTIVITY ❸

Three sets of number cards (1 to 9)

Shuffle the cards and place them face down in a pile. Take three cards, and make a 3-digit number, e.g. 572. Add 31 and write the addition: 572 + 31 = 603.

Repeat until all the cards are used and write the answers in order.

ACTIVITY ❹

A dice

Throw the dice three times to make a 3-digit number, e.g. 646.

Add 49 and write the addition: 646 + 49 = 695.

Repeat ten times. Which answer is closest to 500?

Write the answers from smallest to largest in order.

ACTIVITY ❺

A dice

Write 100, and throw the dice, e.g. 5. Make a 2-digit number using the dice throw as tens, i.e. 50.

Subtract this from 1000, and write the answer.

Continue this process, throwing the dice again and subtracting that number of tens from the number left previously, e.g. throwing a 4, making 40 and subtracting 40 from 950.

Keep going until the number is too small to continue. How many throws did it take?

Photocopy Masters

Numeracy Support Book
pages 28 to 31

Associated units

Abacus 6 N11 Addition/subtraction

③ Decimal numbers

Skills

There are two aspects of decimal numbers which children commonly find difficult. Firstly they fail to understand the value of each digit, i.e. that the first digit after the decimal point is the tenths, the second is the hundredths, and so on. Secondly, they fail to appreciate that decimal numbers can be counted along a number line in the same way as other numbers.

Step 1 Use a 100-point number line numbered from 0 to 10. Count along 0·1, 0·2, 0·3, ... 9·8, 9·9, 10.

Step 2 Use a 100-point number line numbered from 0 to 1. Count along 0·01, 0·02, 0·03, ... 0·98, 0·99, 1·00.

Step 3 Discuss the fact that 0·1 is the same as 0·10, and 0·2 is the same as 0·20, etc.

Step 4 Count along a number line, from 0·01 to 0·99. Then continue past 1·00 to 1·01, 1·02, 1·03 ...

Step 5 Use a place-value grid with pounds and pence. Write numbers on the grid and discuss the value of each digit.

Step 6 Match the numbers on the place-value grid to those on the number line. Show that 0·5 differs from 0·05, and so 5p is £0·05 not £0·5.

Tactics

Number lines

The number line is an essential image for children to develop. They should be able to visualise a line of numbers appropriate to the context. They should be able to count along this line of numbers just as they can count along the line of whole numbers from 1 to 100. Number lines are so important that it is best if children create their own, either writing the numbers onto blank number lines, or using cards if they find writing difficult.

Money

Children find money a very concrete way to make sense of decimals. £1, 10p and 1p coins (use real coins if possible) are particularly good for helping children to order decimals. Money is relevant outside the classroom and this helps children to compare amounts accurately. The coins can be used in conjunction with a place-value grid (see below), and will help children appreciate the difference between decimals such as 1·03 and 1·30.

Place-value grid

Writing numbers in a place-value grid can help children to recognise the **value** of each digit in its **place**. The chart is most helpful if used in conjunction with coins (see above), and also when emphasising the symmetry about the units place (hundreds – hundredths, tens – tenths, etc.).

hundreds	tens	units		tenths	hundredths
3	5	9	●	2	7

Place-value cards

The place-value cards are a resource specifically designed to emphasise the value of each digit in its place. Children can use these cards to create numbers in the same way that they create 4-digit numbers (using thousands, hundreds, tens and units).

The children start with the whole numbers (say this loudly: *two*), then move onto the tenths (quietly: *point eight*), and finally the hundredths (whispered: *three*). 2·83: *two point eight three*.

Activities

Note: All the activities can be undertaken either by individuals or by small groups. The activities are given here in increasing order of difficulty.

ACTIVITY ❶

Number cards (0 to 9)

Spread out the cards face up.

Make five different 1-place decimal numbers using all 10 cards. Write them in order.

Repeat, making different numbers.

ACTIVITY ❷

Coins (1p, 10p and £1)

Take a handful of coins at random, e.g. six 1p coins, three 10p coins, three £1 coins. Write the total amount in pounds: £3·36.

Repeat five times and write the numbers in order.

ACTIVITY ❸

Blank 10-point number lines

Label the left-hand end of one number line 4. Work along the line writing in the decimal numbers: 4·1, 4·2, ... up to 5.

Repeat for different starting numbers.

ACTIVITY ❹

Place-value cards (units, tenths, hundredths)

Shuffle the cards separately and place them face down in three piles.

Take a card from each pile, and make a decimal number. Read it aloud.

Repeat nine more times, to create ten numbers in all.

Order the numbers and write them, from smallest to largest.

ACTIVITY ❺

Coins, £1, 10p, 1ps

Write as many amounts of money between £1 and £2 as you can, that have a zero in either the tenths or in the hundredths position, e.g. £1·03.

Match the decimal number to appropriate coins, i.e. one pound coin and three penny coins.

Write as many numbers as you can (there are 18 altogether).

Photocopy Masters

Numeracy Support Book
pages 32 to 35

Associated units

Abacus 6 N1 Place-value

4 Multiplying by 10 and 100

Skills

Multiplying by 10 or 100 is so useful it is worth spending the time consolidating this with any children who are unsure. Confidence in this strategy will help children to estimate the answers to large multiplications.

Step 1 Rehearse the ×10 table. Ensure that children are able to give immediate answers to any 1-digit number multiplied by 10. Demonstrate how the pattern can be extended for 11×10, 12×10, 13×10, ... 100×10. Discuss what happens when a multiple of 10 is multiplied by 10, e.g. 40×10, 70×10. Point out the number of zeros in the answer.

Step 2 Write out the ×100 table: 1×100, 2×100, 3×100, ... 10×100. Extend this pattern to 11×100, 12×100, 13×100 ... 90×100. Demonstrate what happens when a multiple of 10 is multiplied by 100, e.g. 40×100, 70×100. Discuss the number of zeros in the answer.

Step 3 Write $1 \times 10 = 10$ and $2 \times 10 = 20$. Read both sentences. Write $1.5 \times 10 = 15$. Point out that 1·5 is between 1 and 2, and 15 is between 10 and 20. Write $3 \times 10 = 30$. Then write 2.5×10 and ask the children what they think the answer will be. Repeat for 3.5×10, 4.5×10 ... 9.5×10.

Step 4 Write $1.5 \times 10 = 15$ and read this together. Then write 1.6×10. Show the children that the answer is 16. Ask them to suggest what 1.7×10 will be. Repeat for 1.3×10, and other 1-place decimal numbers.

Step 5 Write 1×100 and 2×100. Use these facts to help find the answer to 1.5×100. Use 2×100 and 3×100 to help demonstrate that $2.5 \times 100 = 250$.

Step 6 Show that multiplying by 100 is the same as multiplying by 10 twice, e.g. $1.5 \times 10 = 15$ and $15 \times 10 = 150$.

Tactics

Place-value cards

The place-value cards can be very useful for demonstrating exactly what happens to each digit when multiplying by 10 or 100. Place the cards in sets: thousands, hundreds, tens and units. Start with a tens card, e.g. 40. Multiply this by 100. Write down the answer: 4000. Find the matching thousands card and compare with the original. How many more zeros does the answer have? How many zeros does 100 have? Repeat this with other place-value cards.

Extend to other 2-digit numbers, multiplying by 10 or 100. Start with a number like 23 (made from 20 and 3). Multiply by 100 and find cards to match the answer: 2000 and 300. Compare each card with those used in the original number. Each one has two more zeros.

Calculator

A calculator is a useful tool which can help children recognise the patterns involved in multiplying by 10 or 100, by removing the arithmetical difficulties. Ask the children to perform a series of multiplications involving 1-place decimal numbers using the calculator. E.g. 3.6×10, 4.6×10, 6.6×10, etc. Encourage them to estimate the answer before pressing '='.

Activities

Note: All the activities can be undertaken either by individuals or by small groups. The activities are given here in increasing order of difficulty.

ACTIVITY ❶ **Number grid (110 to 990), number cards (11 to 99)**
Shuffle the cards and place them face down in a pile.

Choose a card and multiply it by 10. Colour the answer on the grid.

Continue for each card, and then list all the coloured numbers in order.

ACTIVITY ❷ **Tens cards (10, 20, 30, ... 100), post-it notes**
Shuffle the cards and place them face down in a pile.

Take a card, multiply it by 10 and write the answer on a post-it note. Stick it on the number card.

Continue for each card, arranging them in order and writing a list of all the multiplications: $10 \times 10 = 100$, $20 \times 10 = 200$, ...

ACTIVITY ❸ **Number track (0, 100, 200, 300, ... 2000), a coin, a counter**
Place the counter on 0 and spin the coin. If it lands heads – move one space, if it lands tails – move two spaces. Multiply the number you land on by 10 and write the multiplication.

Continue to the end of the track, and repeat, this time multiplying by 100.

ACTIVITY ❹ **A dice, a calculator**
Throw the dice twice, and write down a 2-digit number – the first throw for the tens, the second for the units, e.g. 2 and 6 gives 26.

Multiply this number by 100 and write down the answer.

Using the calculator multiply the number by 10 twice, recording each answer. Is the final answer on the calculator the same as your answer? Repeat.

ACTIVITY ❺ **Two sets of cards (1–9), post-it notes**
Shuffle the cards and place them in a pile face down.

Take two cards and use the numbers to create a 1-place decimal number. E.g. 4·8. Multiply this number by 10, write the answer on a post-it note and stick this on the cards.

Continue until all the cards have been taken.

ACTIVITY ❻ **A number line (1 to 3, marked in tenths), a blank 20-division number line**
Look at the first number on the line. Multiply it by 100 and write the answer in the matching place on the blank number line.

Continue, multiplying each number on the line by 100 and writing the answer in the matching place on the blank line.

Photocopy Masters
Numeracy Support Book
pages 36 to 39

Associated units
Abacus 6 N15 Place-value

⑤ Adding to 10 and adding to 1

Skills

Children who have not learned the addition bonds to 10 by the age of ten need some fundamental teaching in order to remedy this. More of the same type of practice is unlikely to help. Rather than remaining static, we move the children on by adding pairs of single place decimal numbers to make 1. This will reinforce the bonds to 10 as well as increasing the situations in which they can be used. It is crucial to use real 10p coins.

Step 1 Start with 50p + 50p = £1. Use this as the 'base fact' from which all the others follow. Write these numbers as decimals, $0{\cdot}5 + 0{\cdot}5 = 1$. Match each decimal to that number of 10p coins. Stress that five 10p coins is $0{\cdot}5$ of a pound.

Step 2 Move on to $1 + 0$, $0{\cdot}9 + 0{\cdot}1$, and $0{\cdot}8 + 0{\cdot}2$. Use 10p coins to reinforce these facts, starting with the idea that we add 0 to 1 to get 1. We must add point one to point nine to make one, point two to point eight, point seven, … . Match each decimal to that number of 10p coins.

Step 3 Use knowledge of addition bonds to enable children to recognise how many must be added to make the next pound. So $2{\cdot}4$ and $0{\cdot}6$ is 3. First find the next number of pounds (£3) and then focus on the 10ps ($0{\cdot}4$ or 40p) and how many more are needed to make a whole pound.

Tactics

Beads

You can make or buy sets of beads which are very useful in helping to reinforce the $10 + 0$, $9 + 1$, $8 + 2$, … number bonds.

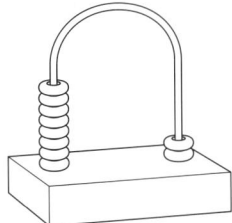

If all the beads are on one side, there are none on the other side ($10 + 0 = 10$). If one bead is moved across, it shows $9 + 1$, and similarly for $8 + 2$, etc. Children enjoy working with these beads and can write down and record the combinations they make.

Fingers

Use fingers to help children see the more difficult facts, e.g. $4 + 6$ and $7 + 3$. Fold down six fingers. *How many are still standing?* Fold down three fingers. *How many are still standing?* Use fingers as a quick reminder of the more difficult facts.

Coins

Use 10p, 50p and £1 coins to help children learn some of the key facts. Use real coins – these can be more effective than the plastic versions.

Draw ten 10p coins and label it as £1. Draw a £1 coin to show the equivalence. Start with 50p and 50p making £1. Write this as a decimal addition, $0{\cdot}5 + 0{\cdot}5 = 1$. Make two piles of 10p coins, and show that adding the two piles makes £1. *Four 10ps and six 10ps make £1.* Write each set as a decimal addition: $0{\cdot}4 + 0{\cdot}6 = 1$. Constantly relate the decimal number, e.g. $0{\cdot}8$ to the number of 10p coins, i.e. eight. Stress that ten 10ps = £1.

Number track

In order to find the next multiple of 10, a number track is very useful. Place a finger or a counter on the track, e.g. on 42. *What is the next multiple of 10? Look at the units. Estimate (don't count) how many spaces.* E.g. for 42, point out that $2 + 8 = 10$, so $42 + 8 = 50$, the next multiple of 10.

Activities

Note: All the activities can be undertaken either by individuals or by small groups. The activities are given here in increasing order of difficulty.

ACTIVITY ❶

Number cards (0 to 10), a blank 6 × 2 grid

Shuffle the cards and spread them out face up on the table.

Make pairs that add to 10 and place them next to each other on the grid: 0 and 10 at the top, then 1 and 9, etc.

When all the cards are laid out, make a copy of the grid.

ACTIVITY ❷

Pegs and a ruler

Clip the pegs onto the ruler, all together at one end.

Make as many different arrangements as you can by moving the pegs along the ruler, e.g. one at one end and nine at the other. Copy each arrangement and write a matching addition.

ACTIVITY ❸

Ten 10p coins, one £1 coin

How many different ways can you make £1 using the 10p coins? Write them all down as decimal additions, e.g. for one 10p and nine 10ps, write $0.1 + 0.9 = 1$. You can also write 10p + 90p = £1. Write five pairs like this and then read them all.

ACTIVITY ❹

Number grid (1 to 100), interlocking cubes

Place red cubes on all the numbers that are 1 less than a multiple of 10. Place pink cubes on all those that are 2 less than a multiple of 10. Place yellow cubes on those that are 3 less than a multiple of 10, etc.

Pick a red number and write the addition to make the next multiple of 10, e.g. $39 + 1 = 40$. Then write a matching addition of 1-place decimal numbers to make the next whole number, i.e. $3.9 + 0.1 = 4$. Repeat for three more red numbers.

Repeat, picking a pink number and writing the matching additions, e.g. $58 + 2 = 60$ and $5.8 + 0.2 = 6$. Do this for four pink numbers.

Repeat this for some yellow numbers, etc.

ACTIVITY ❺

Number track (0 to 100), a dice, a counter, cubes

Place the counter on '0'. Throw the dice and move the counter along the track. Look at the number, e.g. 34. *How many to the nearest multiple of ten?* Take that number of cubes and write down the addition, i.e. $34 + 6 = 40$.

Continue until the end of the track. *How many cubes in total?*

ACTIVITY ❻

Two sets of number cards (1 to 9), post-it notes

Shuffle the cards and place them face down in a pile.

Take two cards. Write these as a 1-place decimal number. *How many to the next whole number?* Write the answer on a post-it note and stick it on the cards. Repeat eight times.

Photocopy Masters

Numeracy Support Book pages 40 to 44

Associated units

Abacus 6 N10 Addition/subtraction

6 Adding several numbers

Skills

Children who try to add a string of 1-digit numbers, without any clear strategies (apart from counting on) will quickly run into difficulties.

Step 1 Encourage the children to recognise pairs of numbers which make 10. They should always scan a series of numbers for these pairs.

Step 2 Start with the largest number, then add any pairs that make 10.

Step 3 Remember that adding 9 is easy – just add 10 and take away 1.

Step 4 Only at the last stage, use counting on with the remaining numbers. The children can put a line under each number as they add to help them remember which numbers remain.

$$\underline{2} + 9 + 17 + \underline{8} + 4 \qquad (2 + 8 = 10)$$
$$\underline{2} + 9 + \underline{17} + \underline{8} + 4 \qquad (17 + 10 = 27)$$
$$\underline{2} + \underline{9} + \underline{17} + \underline{8} + 4 \qquad (27 + 9 = 36)$$
$$\underline{2} + \underline{9} + \underline{17} + \underline{8} + 4 \qquad (36 + 4 = 40)$$

Tactics

This is one of the few occasions in teaching numeracy where a number line is not an appropriate strategy. Children should be discouraged from counting on along the number line starting at the first number.

Addition bonds to 10 grid

Use a simple grid showing the bonds to 10 to remind children to look for pairs of numbers that make 10.

1	and	9	make	10
2	and	8	make	10
3	and	7	make	10
4	and	6	make	10
5	and	5	make	10

Practising adding 9 or 11 using a number grid (1 to 100)

Help children to add 10, by using a number grid (1 to 100), and chanting down the columns, e.g. 24, 34, 44, 54, 64, ...

Now practise adding 9 or adding 11. Make sure that the children add 10 and then adjust. Help them to remember that 9 is 1 **less**, and 11 is 1 **more**.

Fingers

Use fingers **only** if the child has difficulty with counting on in the last stage of the addition.

The strategies above will help to discourage the use of a simplistic counting on strategy. The children should only use their fingers when they have used all the other strategies (i.e. they have looked for pairs that make 10; started with the largest number; added any nines).

Activities

Note: All the activities can be undertaken either by individuals or by small groups. The activities are given here in increasing order of difficulty.

ACTIVITY ❶

Number cards (0 to 10), an extra card '5', stopwatch

Use the cards to find all the possible pairs that make 10.

Reshuffle the cards and repeat. Time yourself. How quickly can you make all the pairs? Try ten times and write down all your times. What is your fastest time?

ACTIVITY ❷

Number cards (10 to 20), two dice

Shuffle the cards and place them face down in a pile. Throw both dice and take a card.

Write the addition of all three numbers, e.g. 3 + 4 + 15. Remember to start with the largest number when completing the addition.

Repeat.

ACTIVITY ❸

Number grid (1 to 100), a counter, a coin

Place the counter on 1. Toss the coin. If it lands heads move one space down. If it lands tails, move one space across. Look at the number you land on. Write it down and add 9.

Repeat until the counter falls off the edge of the grid.

ACTIVITY ❹

Number cards (0 to 9)

Shuffle the cards and place them face down in a pile.

Take five cards, and use them to make a 2-digit number, and three 1-digit numbers. Write them down and add them together, e.g. 42 + 8 + 1 + 9 = 60.

Can you rearrange the cards to make numbers which will give a larger total?

What are the largest and smallest possible totals using these five cards?

Photocopy Masters

Numeracy Support Book
pages 45 to 48

Associated units

Abacus 6 N11 Addition/subtraction

7 Multiplication facts

Skills

Children may have developed a 'block' about multiplication facts, and it is important to counter this 'I can't do it' attitude. This is best achieved by restricting the memorising to a few multiplication facts to begin with, achieving near perfect accuracy in these. The resulting gain in confidence will often allow further progress to be made.

Step 1   Rehearse the ×10 table. Ensure that answers to questions such as: *What are four tens?* and *Six times ten?* are virtually automatic. Ask parallel questions, such as: *How many tens in sixty?* to show the children that they know the answer however the question is asked.

Step 2   Rehearse the ×2 and ×5 tables. Count in twos and in fives up to at least 50. Stress the repetitive elements of these chants. Use fingers to help the children remember how many twos or fives.
E.g. *Six fives are …?* Five (holding up one finger), *ten* (holding up two fingers), *fifteen* (holding up three fingers), and so on.

Step 3   Teach the children the ×9 table on their fingers. (See below.)

Step 4   Rehearse the square numbers, which can be learned by heart as a 'special' sequence.

Step 5   Learn the ×3 table. Tell the children that this is the **only** table they have to memorise.

Step 6   Summarise the position so far: the 1s and 10s are already known; use fingers to find the 2s, 5s and 9s; double ×2 facts to find ×4 facts; double ×3 facts to find ×6 facts; reverse the order to find more facts; the square numbers have been learned. This leaves only one fact to be memorised and that is 7 × 8. Write 56 = 7 × 8 and show the children that the numbers run 5, 6, 7, 8 to help them remember.

Tactics

Tables grid

The main teaching point in helping children who do not know their tables is that there is only one table to be learned: the ×3 table. Demonstrate by asking children to colour the tables on a multiplication grid.

1. They colour the ×1 and the ×10 tables. *You know these.*
2. They colour the ×2, ×5 and ×9 tables. *Use your fingers.*
3. They colour the ×3 table. *This one is special, we have to learn it.*
4. They colour the ×4 and ×6 tables. *Find these by doubling the ×2 and ×3.*
5. They colour any remaining square numbers.
6. They circle the one remaining multiple (56).

Money

Use real 2p, 5p and 10p coins to practise the ×2, ×5 and ×10 tables. Make a line of 2p coins and ask the children to count along it, writing each total: 2p, 4p, 6p … Repeat for 5p and 10p coins.

Number grid (1 to 100)

Colour the ×10 table on the grid and then cross-hatch the ×5 table in a different colour. Colour the ×2 table on a different grid. The children should write down each set of numbers. Draw their attention to the pattern in the units digit to help them remember these numbers.

Interlocking cubes

Use interlocking cubes to build a 2 × 2 square, a 3 × 3 square, a 4 × 4 square and so on. Write the square numbers up to 144.

Activities

Note: All the activities can be undertaken either by individuals or by small groups. The activities are given here in increasing order of difficulty.

ACTIVITY ❶ **Number track (1 to 50).**
Colour all the numbers in the ×2 table red. Colour the numbers in the ×5 table blue. Which numbers are two-colour?

ACTIVITY ❷ **Number cards (1 to 10)**
Shuffle the cards and place them face down in a pile.

Take a card, multiply it by 2, and write the answer. Multiply the card by 5 and write the answer. Multiply the card by 10 and write the answer.

Record your answers in a table.

Card	×2	×5	×10
3	6	15	30

Repeat for each card.

ACTIVITY ❸ **Interlocking cubes**
Build a 2×2 array. *How many cubes?*

Build a 3×3 array. *How many cubes?*

Continue for a 4×4, 5×5, ... 10×10 array.

Write the square numbers in order.

ACTIVITY ❹ **Cassette**
Write down the ×3 table. Write a rhyme for each answer.

Record yourself on a tape saying the ×3 table and also saying your own rhyme for each answer. E.g. *One times three is three, pea. Two times three is six, sticks. Three times three is nine, mine. Four times three is twelve, delve, ...*

ACTIVITY ❺ **A stop watch**
Write the numbers 1 to 10 along the top of a page.

Multiply each one by 9 and write the answers below them. Time yourself. Repeat this several times. What is your best time?

ACTIVITY ❻ **Grid (as shown) drawn on the board**

2	9	4
7	5	3
6	1	8

Add each row and each column. What do you notice?

Make a new grid, multiplying each number by 3. Add each row and each column. What do you notice?

Double each number in this grid to multiply the numbers on the original grid by 6 and make a third grid. Add each row and each column.

Photocopy Masters
Numeracy Support Book pages 49 to 52

Associated units
Abacus 6 N3 Multiplication/division

8 Multiplying decimals

Skills

Step 1 Check that the children are confident with multiplying a multiple of 10 by a 1-digit number, e.g. 7×40. If necessary use place-value cards to help children understand this.

Step 2 Extend this to multiplying 2-digit numbers by a 1-digit number, e.g. 4×38. Use place-value cards to reinforce the method of partitioning, i.e. 4×38 is 4×30 and 4×8.

Step 3 Discuss multiplying a 1-place decimal number between 0 and 1 by a 1-digit number, e.g. 4×0.6. Use 10p coins to demonstrate that the answer is likely to cross over into units (pounds).

Step 4 Extend this to larger 1-place decimal numbers, again using pound and ten pence coins.

Tactics

Place-value cards

These are very useful initially, particularly to stress how to multiply a 2-digit number by partitioning. The cards enable the children to physically partition the number and then perform the two separate multiplications.

Use post-it notes so that the children can write the answers on the cards. Then they add the two parts to obtain the answer. So for 4×63, they use place-value cards 60 and 3, multiply 60 by 4 and write 240 on a post-it note which is stuck on the 60 card, then multiply 3 by 4 and write 12 on the post-it note which is stuck on the 3 card. They then add the two numbers on the post-it notes.

This method can be extended to use decimal place-value cards for multiplying 1-place decimals.

Coins

Coins (£1 and 10p) can be very useful, particularly when multiplying 1-place decimals between 0 and 1. For example, to multiply 0·4 by 6, if we stress that we can represent 0·4 as zero pounds and four 10p coins, we can multiply the four 10p coins by 6. This gives 240p or £2·40. The use of the coins will help the children to see that 0.4×6 gives an answer which is greater than 1. This avoids the common error of $0.4 \times 6 = 0.24$.

Metres and decimetres

Using metre rules and decimetre rods (Dienes or Cuisenaire) is another very visual and tactile way of illustrating that 0.4×6 gives an answer which is greater than 1. If there are several metre rules, this also works for multiplications such as 1.7×3. The children can see that we find $1\,m \times 3$, giving us 3 metre rules, and then 0.7×3, giving us 3 lots of 7 decimetre rods, which is 21 rods. They can then see that this is more than two metres.

Activities

Note: All the activities can be undertaken either by individuals or by small groups. The activities are given here in increasing order of difficulty.

ACTIVITY ❶

Place-value cards (T, U), post-it notes

Shuffle each set of cards and place them in a pile face down.

Take one card from each pile. Multiply each of the numbers by 6, write the total on a post-it note and stick it on the card. Write a multiplication of a 2-digit number, adding the two post-it note numbers to find the answer. E.g. for 30 and 4, $30 \times 6 = 180$, $4 \times 6 = 24$, so $34 \times 6 = 180 + 24 = 204$.

Continue until all the cards have been taken.

ACTIVITY ❷

Place-value cards, (U, t), a 10-sided dice, post-it notes

Shuffle each set of cards and place them in a pile face down.

Take one card from each pile. Throw the dice and multiply each of the numbers by the dice number thrown. Write the totals on post-it notes and stick them to the cards. Write a multiplication of a 1-place decimal number, adding the two post-it note numbers to find the answer. E.g. for 7 and 0·5, and dice throw 4, $4 \times 7 = 28$, $4 \times 0{·}5 = 2$, so $4 \times 7{·}5 = 28 + 2 = 30$.

Continue until all the cards have been taken.

ACTIVITY ❸

Two sets of number cards, (1 to 9), a dice

Shuffle the cards and place them face down in a pile.

Take two number cards from the pile and use these to create a 1-place decimal number, e.g. 3 and 7 to make 3·7. Throw the dice. Multiply the decimal number by the dice number thrown. (If a 1 is thrown, multiply by 10.) Write the parts and then the answer on a post-it note and stick it on the cards. E.g. for a dice throw of 4, write 12 (4×3) and 2.8 ($4 \times 0{·}7$) on the post-it note and then add these and write 14·8.

Repeat until all cards have been taken.

ACTIVITY ❹

Dominoes

Shuffle the dominoes and place them face down on the table. Choose a domino and write it as a 1-place decimal number. E.g. becomes 6·4. Choose a number between 3 and 9 and multiply the domino number by this number. Write the two parts and add them to find the answer. E.g. $5 \times 6{·}4 = (5 \times 6{·}0) + (5 \times 0{·}4) = 32$.

Repeat at least six times, choosing a different domino and a different number to multiply by each time.

ACTIVITY ❺

Coins (10ps and £1s)

Take a handful of coins. Write the amount, e.g. £4·80. Multiply the number by 4, writing the calculation as a decimal multiplication, e.g. $4 \times 4{·}8$. Multiply the units and then the tenths. Add the two parts to get the answer, i.e. $4 \times 4 = 16$, $4 \times 8 = 3{·}2$, so $4 \times 4{·}8 = 16 + 3{·}2 = 19{·}2$.

Now multiply the original amount by 10, i.e. $4{·}8 \times 10 = 48$. Work out the difference between the two answers, i.e. $48 - 19{·}2 = 28{·}8$.

Write the original amount again as a decimal number, e.g. 4·8, and multiply this by 6. This should be the same as the answer to the subtraction.

Repeat this process at least five times.

Photocopy Masters

Numeracy Support Book
pages 53 to 56

Associated units

Abacus 6 N31 Multiplication/division

9 Fractions

Skills

Step 1 Choose an image – pizza, cakes, cookies – and cut out some circles of card to represent them. Find one half, one quarter, three quarters. Find one third and two thirds. Check that the children understand that the fractional parts (e.g. quarters) must all be the same size.

Step 2 Put 'olives' round the pizza or 'chocolate chips' round the cake. Distribute these evenly so that when the pizza or cake is halved or quartered, there are the same number of olives (or chocolate chips) on each slice. E.g. if there are 12 olives, there should be 3 on each quarter, 4 on each third, 6 on each half, and so on.

Step 3 Use pizzas with olives (or cakes with chocolate chips) to lead into finding unit fractions of quantities.

Step 4 Extend this to find $\frac{3}{4}$ of a quantity or $\frac{2}{3}$ of another quantity. If the children are confident, try finding other fractions, such as $\frac{4}{5}$ or $\frac{5}{6}$.

Step 5 Use pizzas or cakes to demonstrate $1\frac{1}{2}$, $2\frac{1}{4}$, … . Stress that we have one whole pizza and half a pizza, or two whole cakes and a quarter of a cake, etc.

Step 6 Extend this to find mixed fractions of quantities, e.g. $1\frac{1}{2}$ pizzas with 16 olives on each pizza, how many olives? Use bags of sweets: $2\frac{3}{4}$ bags, containing 12 sweets each, how many sweets?

Tactics

Pizzas, cakes, cookies

It is very useful to have an image of a whole object which can be divided into equal parts to provide a visual image of a fraction. This can then be related to quantities by placing objects (e.g. olives or chocolate chips) on the pizza or cake in a regular pattern around the edge. Then, when the cake is halved, there is half the quantity of olives, when it is quartered there is one quarter, etc. So 12 olives distributed evenly gives 3 on each quarter, 4 on each third, and so on. This provides a practical means of linking fractions of objects (cakes, pizzas, etc) to fractions of quantities.

Bags of sweets and towers of cubes

When teaching mixed fractions it is essential to have a means of making a quantity into a unit, thus a bag of 12 sweets or a tower of 12 cubes is both a quantity and a unit. Fractions can then be calculated in terms of the number of sweets or cubes. E.g. $1\frac{1}{2}$ bags of sweets, with 12 sweets in a bag, means there are 18 sweets ($1\frac{1}{2} \times 12 = 1 \times 12$ and $\frac{1}{2} \times 12$). Show children how to partition the calculation in the same way as partitioning the multiplication of decimals in the previous unit.

Number line

Placing fractions in order along a number line is also very important. As well as understanding what $\frac{1}{4}$ or $1\frac{1}{2}$ are in terms of fractions of a pizza, children also need to understand that these numbers can be placed along a number line like any other number. This will also help them to recognise that they have decimal equivalents, e.g. $\frac{1}{2} = 0\cdot5$, etc. It is important to reinforce this ordinal (position in sequence) aspect of fractional numbers as well as stressing the cardinal (quantity) aspect.

Activities

Note: All the activities can be undertaken either by individuals or by small groups. The activities are given here in increasing order of difficulty.

ACTIVITY ❶ **Interlocking cubes, fraction cards ($\frac{1}{4}$, $\frac{1}{2}$, $\frac{3}{4}$, $1\frac{1}{4}$, $1\frac{1}{2}$, $1\frac{3}{4}$, $2\frac{1}{4}$, $2\frac{1}{2}$, $2\frac{3}{4}$)**
Make towers of twelve cubes. Take a card. Match it to the appropriate number of towers and cubes. Write the matching number sentence for each $1\frac{1}{2}$ towers = 18 cubes.

ACTIVITY ❷ **Fraction cards ($\frac{1}{4}$, $\frac{1}{2}$, $\frac{3}{4}$, $1\frac{1}{4}$, $1\frac{1}{2}$, $1\frac{3}{4}$, $2\frac{1}{4}$, $2\frac{1}{2}$, $2\frac{3}{4}$, ... 4), a dice (numbered 2, 3, 4, 6, 8, 0), post-it notes**
Spread out the cards face up.

Choose a fraction card. Throw the dice twice and create a 2-digit number. This is the number of sweets in a bag. Draw the bag on the post-it note and write the number on it.

Find the fraction on the card you chose of the number on the post-it note. E.g. for 24 and the fraction $2\frac{1}{4}$: *There are twenty-four sweets in the bag and I have two and a quarter bags.* Write the number of sweets, e.g. $2 \times 24 = 48$ and $\frac{1}{4} \times 24 = 6$, so $48 + 6$ is 54. *There are 54 sweets.*

Repeat until all the cards have been used.

ACTIVITY ❸ **Coins (£1, 50p, 20p, 10p, 5p), fraction cards ($\frac{1}{4}$, $\frac{1}{2}$, $\frac{3}{4}$, $1\frac{1}{4}$, $1\frac{1}{2}$, $1\frac{3}{4}$, $2\frac{1}{4}$, $2\frac{1}{2}$, $2\frac{3}{4}$ ... 4)**
Match the fraction cards to coins, e.g. for $1\frac{1}{4}$ take one pound coin, one twenty pence coin and one five pence coin. Write the appropriate number sentence: $1\frac{1}{4}$ pounds = £1·25.

ACTIVITY ❹ **Number line (0 to 3, marked 0·1, 0·2, 0·3 ...), fraction cards ($\frac{1}{4}$, $\frac{1}{2}$, $\frac{3}{4}$, $1\frac{1}{4}$, $1\frac{1}{2}$, $1\frac{3}{4}$, $2\frac{1}{4}$, $2\frac{1}{2}$, $2\frac{3}{4}$)**
Position the fraction cards along the number line. Write all the pairs of equivalences, e.g. $\frac{1}{4}$ = 0·25, $\frac{1}{2}$ = 0·5, and so on.

ACTIVITY ❺ **Coins (£2, £1, 50p, 20p, 10p, 5p, 2p, 1p)**
Take each coin. Draw round it. Write the fraction that coin is of one pound, e.g. 50p is $\frac{1}{2}$ of £1.

Repeat, taking different combinations of coins.

Photocopy Masters
Numeracy Support Book
pages 57 to 60

Associated units
Abacus 6 N7 Fractions/decimals

⑩ Percentages

Skills

Step 1 Rehearse multiplying and dividing by 10. Remind the children that 27 × 10 = 270, and 2·7 × 10 = 27, so 270 ÷ 10 = 27 and 27 ÷ 10 = 2·7. Practise both multiplying and dividing 2- and 3-digit numbers by 10.

Step 2 Use a 2 × 5 × 10 cuboid made of interlocking cubes. Count the number of cubes, to demonstrate that there are 100. Explain that 'per cent' means 'out of one hundred', and introduce the % sign. Relate 'cent' to the US coins (cents), pointing out that there are one hundred cents in a dollar.

Step 3 Remove one of the rows of 10 cubes from the cuboid. Demonstrate that this is ten cubes out of one hundred. *We call this ten percent.* Write this as 10%.

Step 4 Check how many rows of 10 there are. Remove one at a time, showing 20% or twenty out of one hundred, and then 30% or thirty out of one hundred. Continue up to 100%.

Step 5 Find 10% of numbers other than 100. Make a cuboid of 60 cubes (6 × 5 × 2) and remove a row of 6, demonstrating that this is six out of sixty. *There are ten such rows. So this is one tenth or ten percent.* Match this to the equivalent sentence, 10% of 60 = 6.

Step 6 Use finding 10% to help children to find 20%, 30%, etc.

Tactics

Cubes

Use 100 cubes to make the idea of percentages concrete. Build these into a cuboid. Find one tenth or 10%. Find 20% and 50%. Relate these to 10%. Then make different-sized cuboids, using different numbers of cubes, e.g. 60 cubes or 30 cubes or 120 cubes. Always use one tenth as the length of one side of the cuboid. E.g. 30 cubes should be built into a cuboid having a side of 3, 60 cubes should be built into a cuboid having a side of 6. This way, 10% is easy to find. Find 10% or one tenth of these amounts. This is difficult for children. Since 10% of 100 is 10, they may assume that 10% is always 10. Stress that 10% is the same as one tenth. Using different cuboids helps.

Grids

Colour 10% on a 10 by 10 grid. Do this in lots of different ways to demonstrate that 10% does not have to be a row. As long as one tenth of the squares are coloured, then 10% of the grid is coloured. Extend this idea using different-sized grids, e.g. 6 × 5. Count the squares. There are 30. *How many squares are there in ten per cent or one tenth? Three.* Colour three squares in several different ways on 6 × 5 grids. Repeat with different grids.

Calculator

The children can use the calculator to help them work out 10% of several different amounts. Concentrate on multiples of 10 at first, e.g. finding 10% of 30, 40, 130, 260, 380, ... They should write down each calculation, e.g. 10% of 380 = 38. When the children spot the pattern, they can write some more calculations by themselves without using the calculator.

Repeat this process, using the calculator to help the children find 10% of 2-digit numbers, e.g. 37. Do this using the calculator until the children again spot the pattern.

Activities

Note: All the activities can be undertaken either by individuals or by small groups. The activities are given here in increasing order of difficulty.

ACTIVITY ❶

Cubes, a ten-sided dice

Throw the dice. Write the number and multiply it by 10. Take that number of cubes. Find 10% of this number. Write a matching calculation, e.g. 10% of 50 = 5.

Continue throwing the dice and finding 10% of the multiple of 10. If you throw a number which has already been thrown, find 20% of that number. If you have already done this, find 30% and so on.

ACTIVITY ❷

Coins (£1, 10p, 20p, 50p)

Take a handful of coins. Work out how much you have, e.g. two £1s, three 50ps, four 20ps and two 10ps to make £4·50. Write down the amount and draw the coins. Find 10% of this amount and draw the coins.

ACTIVITY ❸

A calculator, post-it notes, number cards (10 to 100)

Shuffle the number cards and place them in a pile face down.

Take a card. Use the calculator and write 10% of the number on a post-it note. E.g. for 47, write 4·7. Repeat ten times, using the calculator. Do ten more without the calculator.

ACTIVITY ❹

Place-value cards (H, T)

Take two cards, one from each pile to make a 3-digit multiple of 10. Write 10% of that number on a post-it note and stick it on the cards. Find another percentage, e.g 20%, and write this. Repeat until all the cards have been taken.

ACTIVITY ❺

Coins (50p, 20p, 10p, 5p, 2p, 1p), (as shown) drawn on the board grid

Start	£1	£1·20	
£2·50	£3·10	£1·70	
£2·80	£5·60	£4·40	End

Place your counter on 'start'. Move your counter around the grid, moving one space at a time either horizontally or vertically, but not diagonally. Each time you land on an amount, find 10% of it and take that number of coins. How much do you have when you reach the 'End'?

Photocopy Masters

Numeracy Support Book pages 61 to 64

Associated units

Abacus 6 N35 Percentages

Numbers to 1000

Write the total for each set.

700 80 4
784

800 70 6
[ ]

900 90 9
[ ]

100 90 1
[ ]

200 10
[ ]

300 3
[ ]

400 10 4
[ ]

800 80
[ ]

600 3
[ ]

300 90 1
[ ]

500 60 5
[ ]

Use number cards 1, 5, 9.

Make as many 3-digit numbers as you can.

Write them down.

Abacus Ginn and Company 2001. Copying permitted for purchasing school only. This material is not copyright free.

Numbers to 10 000

Write the position of each pointer.

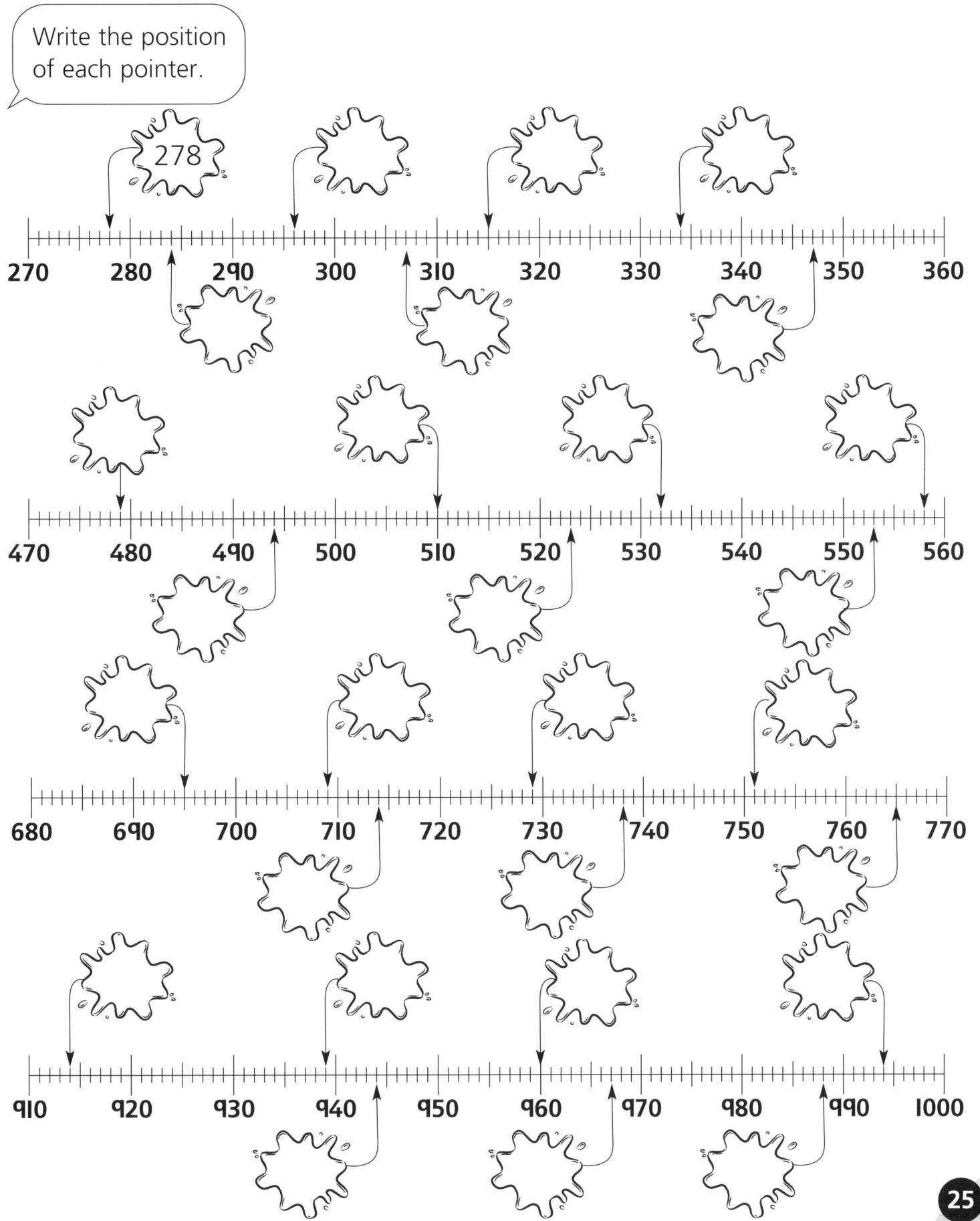

Abacus Ginn and Company 2001. Copying permitted for purchasing school only. This material is not copyright free.

Name _____

Ordering numbers

Write < or > between each pair.

390 **<** 410		612 ☐ 621
7799 ☐ 8002	1311 ☐ 1131	3684 ☐ 3486
1101 ☐ 1110	4287 ☐ 4278	9666 ☐ 9660
1190 ☐ 9109	4210 ☐ 4120	7111 ☐ 7109
13 470 ☐ 17 431	19 014 ☐ 11 409	16 301 ☐ 16 103
47 410 ☐ 47 401	11 010 ☐ 11 000	90 090 ☐ 90 190

Write the numbers on the track in order, from smallest to largest.

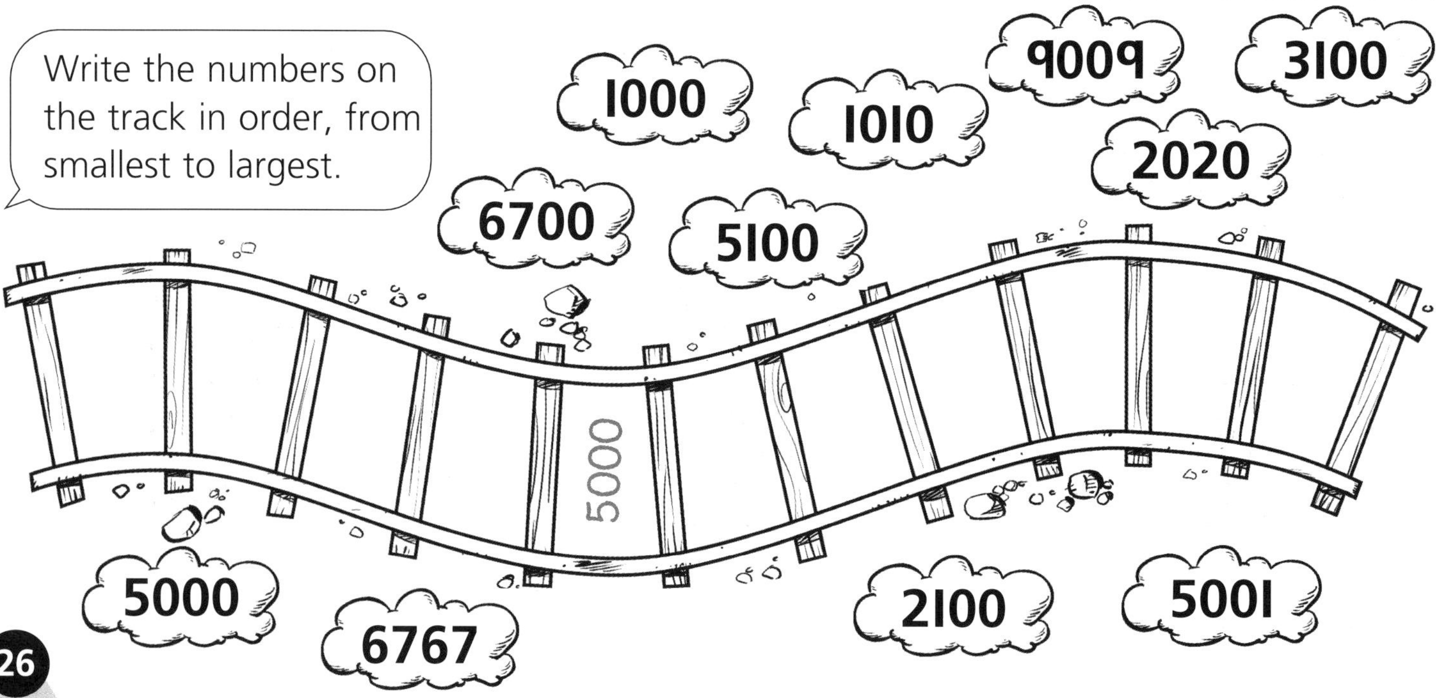

1000　1010　9009　3100　2020

6700　5100

5000

5000　6767　2100　5001

Abacus Ginn and Company 2001. Copying permitted for purchasing school only. This material is not copyright free.

Name ————————————————

Numbers to 100 000

> Write the missing numbers.

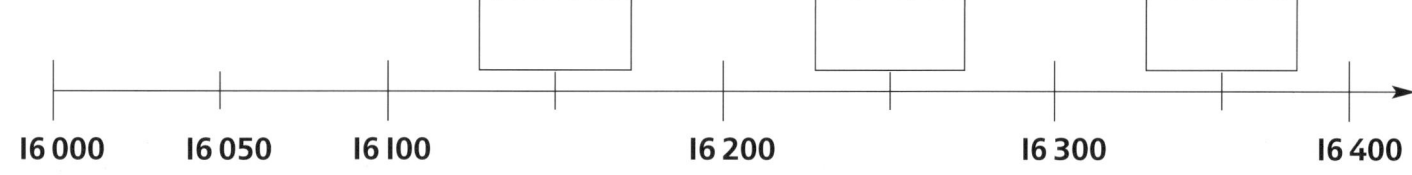

16 000   16 050   16 100   16 200   16 300   16 400

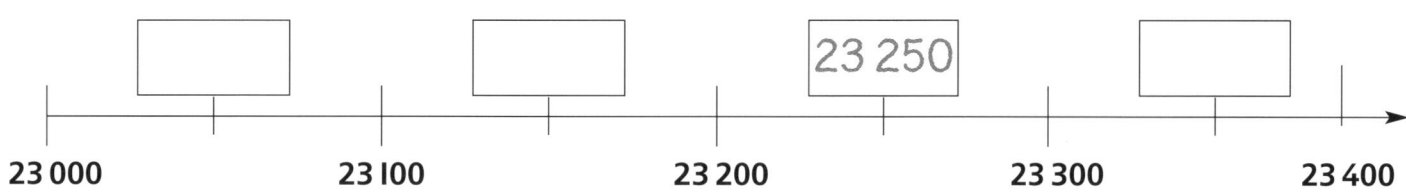

23 000   23 100   23 200   23 300   23 400

Box: 23 250

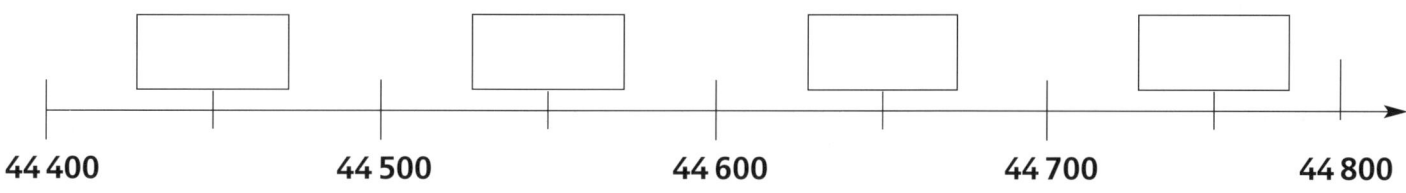

44 400   44 500   44 600   44 700   44 800

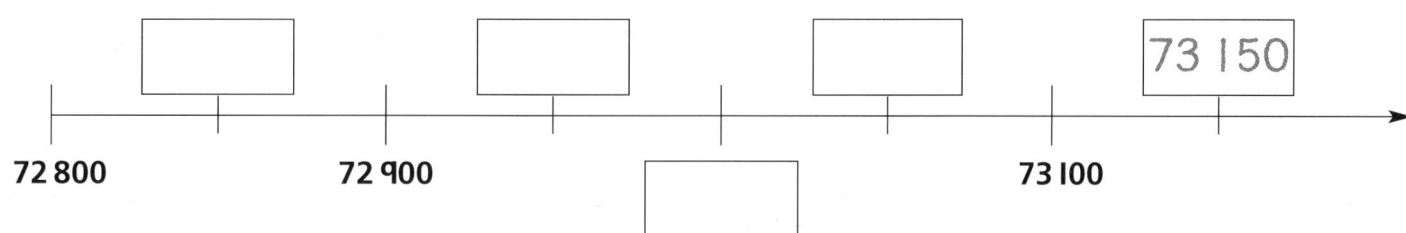

72 800   72 900   73 100

Box: 73 150

> Add the numbers.

16 000	700	7
400	9	11 000
50	24 000	10
+ 7	+ 30	+ 400
————	————	————

Abacus Ginn and Company 2001. Copying permitted for purchasing school only. This material is not copyright free.

Name —————————————————

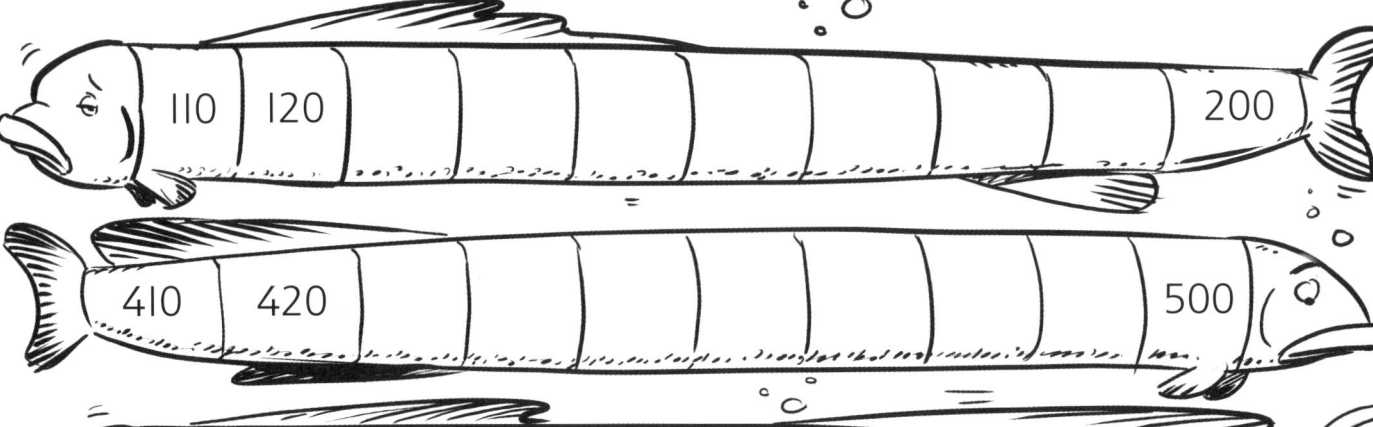

Adding 10

Write the missing numbers.

| 110 | 120 | | | | | | | 200 |

| 410 | 420 | | | | | | | 500 |

| 700 | 710 | | | | | | | |

| 250 | 260 | 270 | | | | | | |

| 540 | 550 | | | | | | | |

| 870 | 880 | | | | | | | |

| 630 | 640 | | | | | | | |

| 380 | 390 | | | | | | | |

Abacus Ginn and Company 2001. Copying permitted for purchasing school only. This material is not copyright free.

Name —————————————

Adding

Complete these addition chains.

164   +   21   =   185

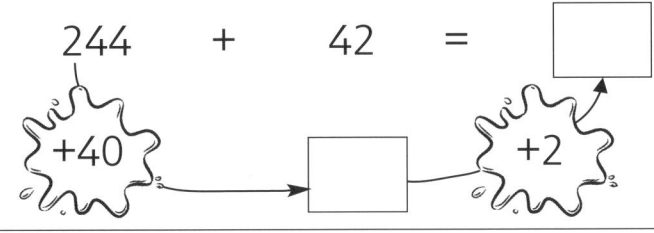
+20   →   184   +1

358   +   31   =   ☐

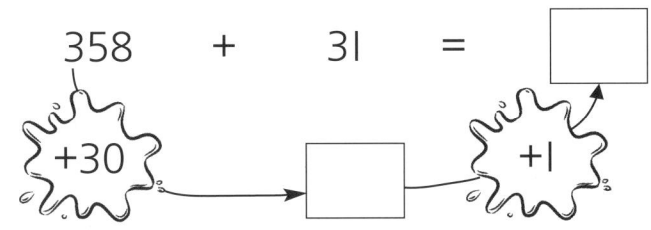
+30   →   ☐   +1

244   +   42   =   ☐

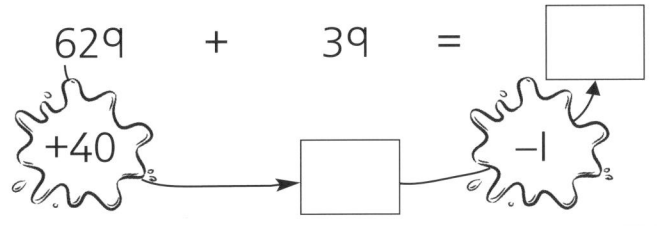
+40   →   ☐   +2

139   +   32   =   ☐

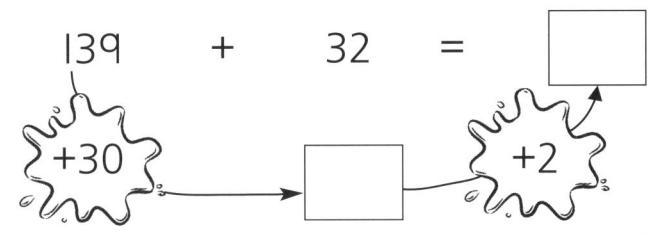
+30   →   ☐   +2

629   +   39   =   ☐

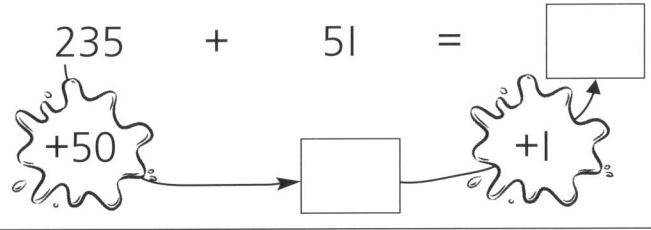
+40   →   ☐   −1

447   +   29   =   ☐

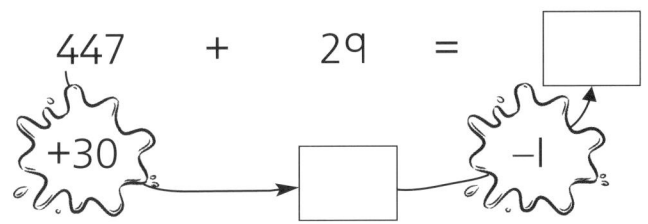
+30   →   ☐   −1

235   +   51   =   ☐

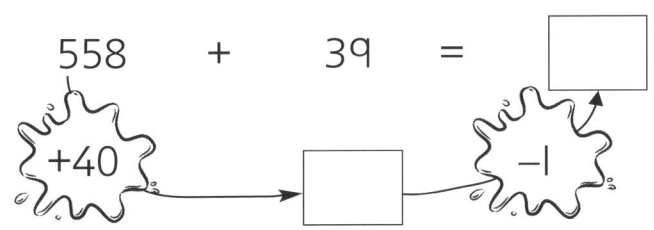
+50   →   ☐   +1

149   +   31   =   ☐

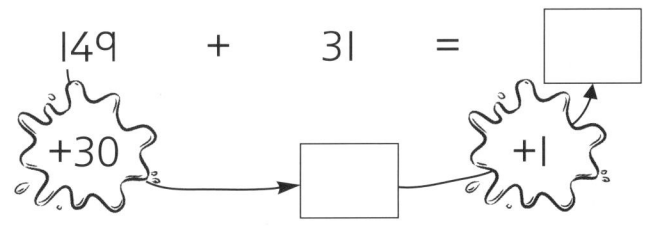
+30   →   ☐   +1

558   +   39   =   ☐

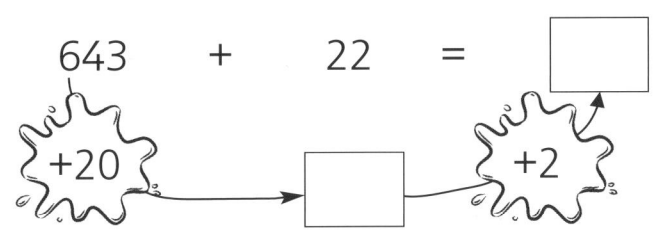
+40   →   ☐   −1

838   +   19   =   ☐

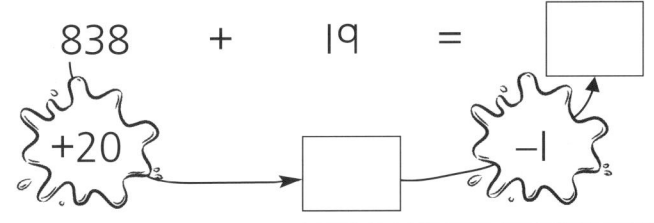
+20   →   ☐   −1

643   +   22   =   ☐

+20   →   ☐   +2

167   +   19   =   ☐

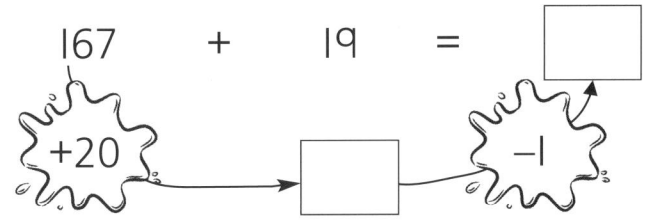
+20   →   ☐   −1

Abacus Ginn and Company 2001. Copying permitted for purchasing school only. This material is not copyright free.

Name _____

Adding

Write the
missing numbers.

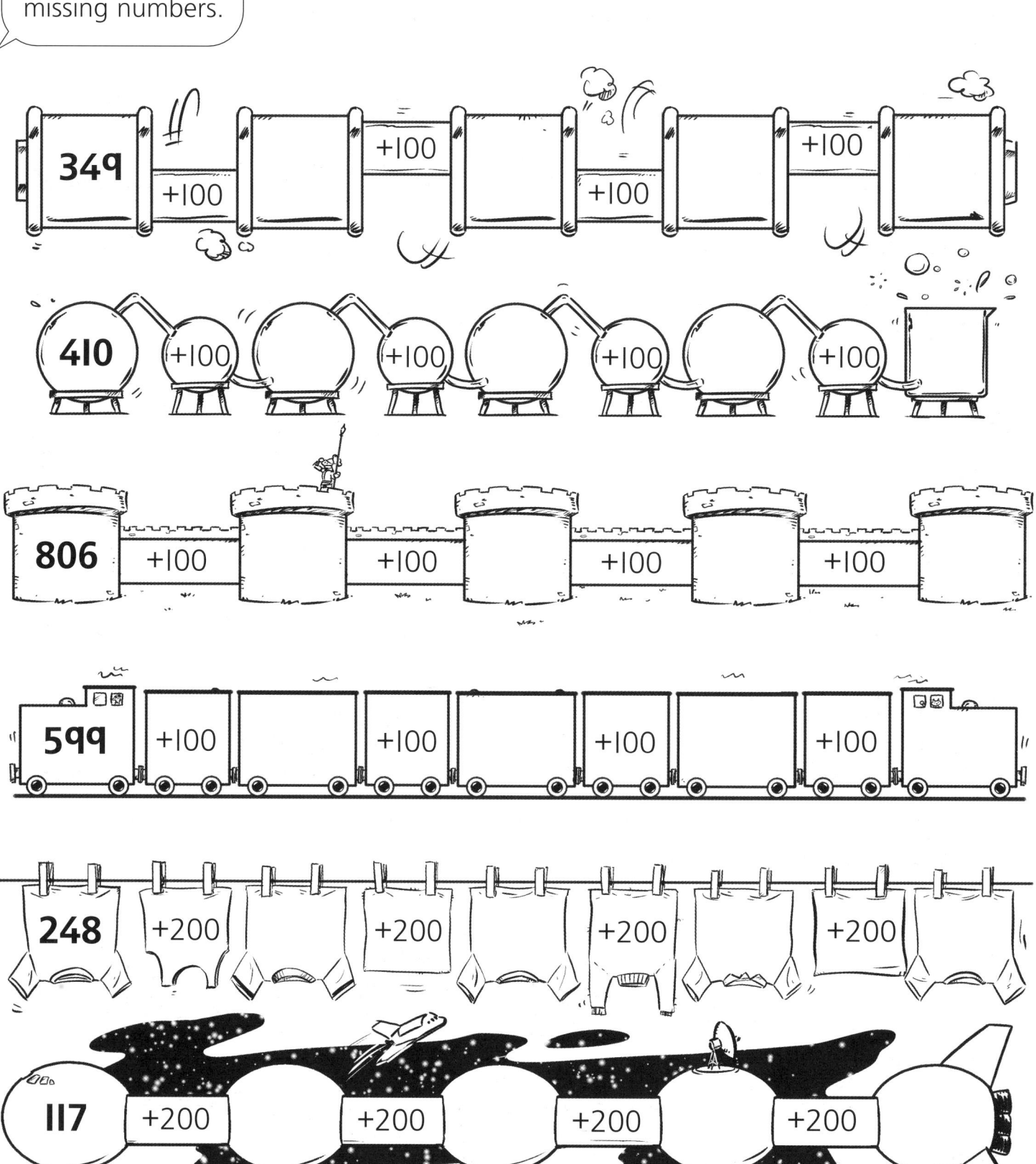

349 +100 +100 +100 +100

410 +100 +100 +100 +100

806 +100 +100 +100 +100

599 +100 +100 +100 +100

248 +200 +200 +200 +200

117 +200 +200 +200 +200

Abacus Ginn and Company 2001. Copying permitted for purchasing school only. This material is not copyright free.

Name —————————————

Adding and subtracting

Write the
number:

 100 less than 290

 190

 100 less than 470

 100 more than 570

 110 more than 260

 120 less than 334

 240 more than 310

 210 less than 410

 320 more than 470

 250 less than 270

 190 more than 330

 140 less than 940

Write the output for each number.

451
287
154
346

+100   −10   +200   −30

Abacus Ginn and Company 2001. Copying permitted for purchasing school only. This material is not copyright free.

Name _____

Decimals

Write the next
two numbers.

| 4·8 | 4·9 | 5·0 | 2·7 | | |

| 3·9 | | | 0·8 | | | 2·9 | | |

| 0·1 | | | 10·9 | | | 4·3 | | |

| 5·5 | | | 1·8 | | | 11·2 | | |

| 0·4 | | | 6·6 | | | 5·9 | | |

Write the
missing numbers.

0·1 0·2 0·3

0·8

1·2

2·3

1·6

Abacus Ginn and Company 2001. Copying permitted for purchasing school only. This material is not copyright free.

Decimals

Write < or >
between each pair.

2·8 $\boxed{<}$ 3·9   1·8 $\boxed{\phantom{<}}$ 1·3

0·8 $\boxed{\phantom{<}}$ 2·1   1·9 $\boxed{\phantom{<}}$ 0·6   1·1 $\boxed{\phantom{<}}$ 0·9

2·1 $\boxed{\phantom{<}}$ 1·2   1·3 $\boxed{\phantom{<}}$ 3·1   2·3 $\boxed{\phantom{<}}$ 2·2

4·1 $\boxed{\phantom{<}}$ 4·4   3·9 $\boxed{\phantom{<}}$ 4·1   5·5 $\boxed{\phantom{<}}$ 4·9

1·9 $\boxed{\phantom{<}}$ 2·2   0·8 $\boxed{\phantom{<}}$ 1·1   10·1 $\boxed{\phantom{<}}$ 9·9

8·8 $\boxed{\phantom{<}}$ 7·7   0·3 $\boxed{\phantom{<}}$ 2·0   6·5 $\boxed{\phantom{<}}$ 5·6

Use number cards
2, 4, 5, 7, 8.

Make as many
different decimal
numbers as you can.

Write them down.

2   **4**   **5**   **7**   **8**

33

Abacus Ginn and Company 2001. Copying permitted for purchasing school only. This material is not copyright free.

Name ——————————————

Decimals

Write the position of each balloon.

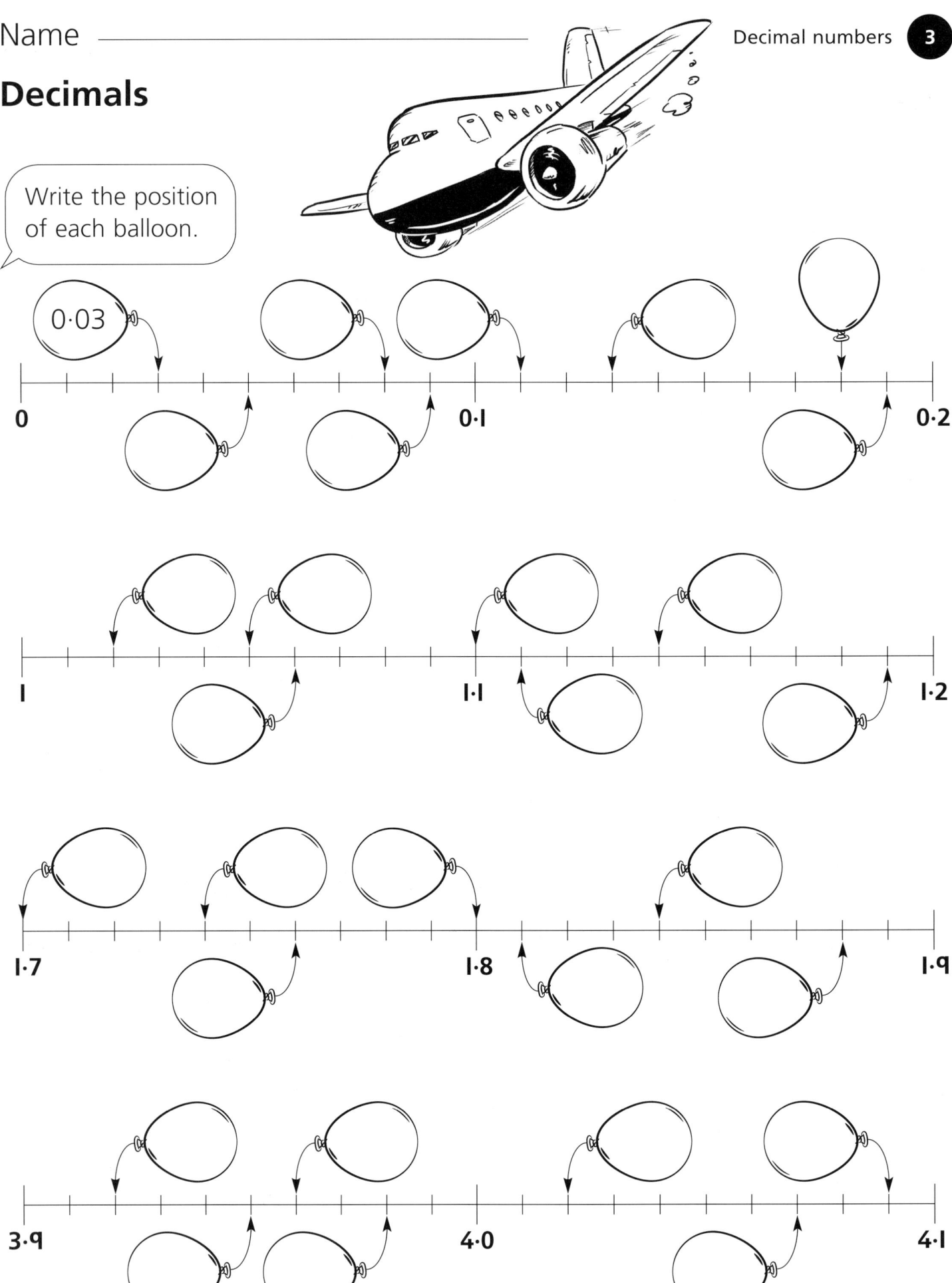

Abacus Ginn and Company 2001. Copying permitted for purchasing school only. This material is not copyright free.

Name ———————————————

Decimals

Write how
many pounds.

£ | 1·32 |

£ [   ]

£ [   ]

£ [   ]

£ [   ]

£ [   ]

£ [   ]

£ [   ]

£ [   ]

£ [   ]

£ [   ]

£ [   ]

£ [   ]

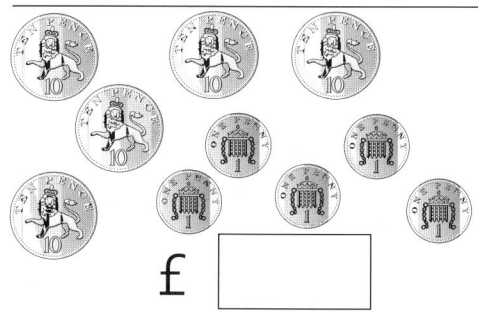

£ [   ]

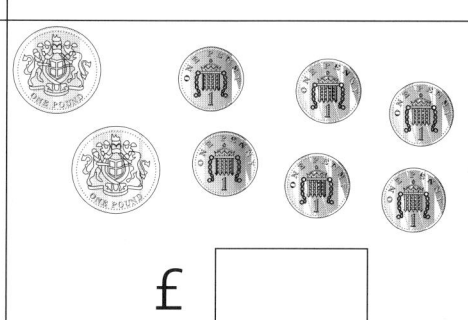

£ [   ]

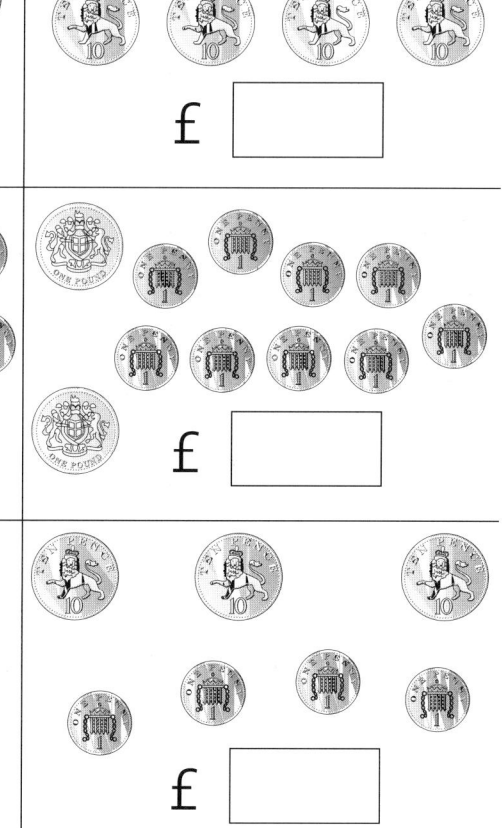

Abacus Ginn and Company 2001. Copying permitted for purchasing school only. This material is not copyright free.

Name _____

Multiplying by 10

Complete the multiplication table, using a calculator.

	14	15	16	17	18		20	21	22	23		25	26	27	28	
×10	140				190				240						290	

Complete these multiplications.

10 × 18 =

10 × 29 =

10 × 38 =

10 × 31 =

10 × 40 =

10 × 35 =

10 × 49 =

10 × 43 =

10 × 50 =

10 × 58 =

10 × 52 =

10 × 60 =

10 × 64 =

10 × 47 =

10 × 53 =

10 × 30 =

10 × 72 =

36

Abacus Ginn and Company 2001. Copying permitted for purchasing school only. This material is not copyright free.

Name —————————————————————

Multiplying by 10 and 100

Complete the
multiplication table.

	2	2·5	3	3·5	4	4·5	5	5·6	5·9	6
×10	20									

Put a loop around
the correct answer.

10 × 4·8	408 480 (48)	10 × 2·4	24 214 204
10 × 5·5	505 515 55	10 × 4·4	404 44 40·4	10 × 6·6	66 60·6 166
10 × 7·1	701 71 171	10 × 2·9	290 291 29	10 × 9·9	909 99·0 919
10 × 10·1	101 110 100·1	10 × 2·8	28·0 280 28·8	10 × 4·7	47 407 470
100 × 9·8	98 980 9800	100 × 1·1	111 110 1101	100 × 5·5	55 55·0 550

Abacus Ginn and Company 2001. Copying permitted for purchasing school only. This material is not copyright free.

Multiplying by 10 and 100

Write how much 10 of each magazine cost.

$10 \times £3·30 = $ £33

$10 \times £1·20 = $ ☐

$10 \times £2·40 = $ ☐

$10 \times £4·25 = $ ☐

$10 \times £1·75 = $ ☐

$10 \times £0·85 = $ ☐

Write the missing numbers.

Abacus Ginn and Company 2001. Copying permitted for purchasing school only. This material is not copyright free.

Name ————————————

Multiplying by 10 and 100

$10 \times 0.3 = 3$

$10 \times 3 = 30$

0·3 cm $100 \times 0.3\ \text{cm} = 30\ \text{cm}$

Each giant insect grows to be 100 times larger.

0·4 cm

$10 \times 0.4 = \underline{\hspace{1cm}}$

$10 \times \underline{\hspace{1cm}} = \underline{\hspace{1cm}}$

$100 \times 0.4\ \text{cm} = \boxed{}$ cm

0·9 cm

$10 \times 0.9 = \underline{\hspace{1cm}}$

$10 \times \underline{\hspace{1cm}} = \underline{\hspace{1cm}}$

$100 \times 0.9\ \text{cm} = \boxed{}$ cm

0·7 cm

$100 \times 0.7\ \text{cm} = \boxed{}$ cm

0·5 cm

$100 \times 0.5\ \text{cm} = \boxed{}$ cm

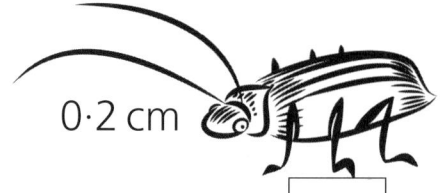

0·2 cm

$100 \times 0.2\ \text{cm} = \boxed{}$ cm

1·2 cm

$100 \times 1.2\ \text{cm} = \boxed{}$ cm

1·5 cm

$100 \times 1.5\ \text{cm} = \boxed{}$ cm

0·1 cm

$100 \times 0.1\ \text{cm} = \boxed{}$ cm

Multiply each balloon number by 10.

Then multiply by 100.

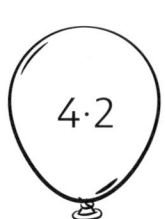

4·2

9·8

3·6

7·4

6·1

$3.6 \times 10 = \underline{\hspace{1cm}}$

$3.6 \times 100 = \underline{\hspace{1cm}}$

$7.4 \times 10 = \underline{\hspace{1cm}}$

$7.4 \times 100 = \underline{\hspace{1cm}}$

$4.2 \times 10 = \underline{\hspace{1cm}}$

$4.2 \times 100 = \underline{\hspace{1cm}}$

$9.8 \times 10 = \underline{\hspace{1cm}}$

$9.8 \times 100 = \underline{\hspace{1cm}}$

$6.1 \times 10 = \underline{\hspace{1cm}}$

$6.1 \times 100 = \underline{\hspace{1cm}}$

Abacus Ginn and Company 2001. Copying permitted for purchasing school only. This material is not copyright free.

Next ten

Write the next ten. | Use the number grid to help you.

1	2	3	4	5	6	7	8	9	10
11	12	13	14	15	16	17	18	19	20
21	22	23	24	25	26	27	28	29	30
31	32	33	34	35	36	37	38	39	40
41	42	43	44	45	46	47	48	49	50
51	52	53	54	55	56	57	58	59	60
61	62	63	64	65	66	67	68	69	70
71	72	73	74	75	76	77	78	79	80
81	82	83	84	85	86	87	88	89	90
91	92	93	94	95	96	97	98	99	100

24 → ☐   36 → ☐

58 → ☐   28 → ☐

63 → ☐   74 → ☐

91 → ☐   89 → ☐

48 → ☐   31 → ☐

98 → ☐   85 → ☐   42 → ☐   67 → ☐

55 → ☐   77 → ☐   34 → ☐   53 → ☐

Write the next ten.

116 → ☐   234 → ☐

296 → ☐   409 → ☐   338 → ☐   684 → ☐

555 → ☐   921 → ☐   719 → ☐   847 → ☐

Abacus Ginn and Company 2001. Copying permitted for purchasing school only. This material is not copyright free.

Next ten

Write how many
to the next ten.

6
34 → 40

46 → 50

72 → 80

66 → 70

21 → 30

53 → 60

85 → 90

18 → 20

37 → 40

24 → 30

59 → 60

87 → 90

76 → 80

68 → 70

51 → 60

25 → 30

48 → 50

42 → 50

83 → 90

93 → 100

Write how many
to the next ten.

31	32	33	34	35	36	37	38	39
9								
40	40	40	40	40	40	40	40	40

Abacus Ginn and Company 2001. Copying permitted for purchasing school only. This material is not copyright free.

Adding to 1

Write how much to make £1.

90p + ☐ p = 100p

£0·90 + £ ☐ = £1·00

50p + ☐ p = 100p

£0·50 + £ ☐ = £1·00

70p + ☐ p = 100p

£0·70 + £ ☐ = £1·00

20p + ☐ p = 100p

£0·20 + £ ☐ = £1·00

60p + ☐ p = 100p

£0·60 + £ ☐ = £1·00

80p + ☐ p = 100p

£0·80 + £ ☐ = £1·00

10p + ☐ p = 100p

£0·10 + £ ☐ = £1·00

30p + ☐ p = 100p

£0·30 + £ ☐ = £1·00

40p + ☐ p = 100p

£0·40 + £ ☐ = £1·00

100p + ☐ p = 100p

£1·00 + £ ☐ = £1·00

0p + ☐ p = 100p

£0·00 + £ ☐ = £1·00

Complete the table for making 1.

0·1	0·2	0·3	0·4	0·5	0·6	0·7	0·8	0·9	1
and									
0·9									
make 1									

Abacus Ginn and Company 2001. Copying permitted for purchasing school only. This material is not copyright free.

Next whole number

Add to each to make the
next whole number.

1.
$0.8 + 0.\boxed{2} = 1$

2.
$0.6 + 0.\boxed{} = 1$

3.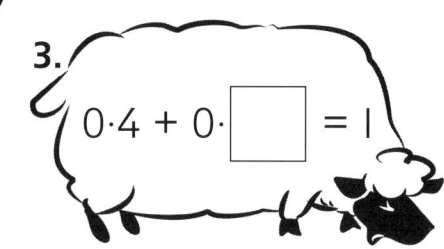
$0.4 + 0.\boxed{} = 1$

4.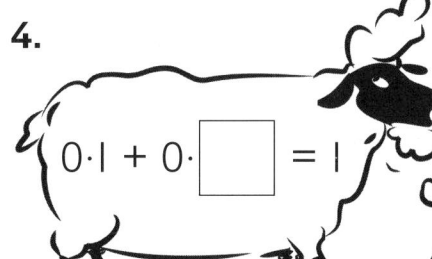
$0.1 + 0.\boxed{} = 1$

5.
$0.7 + 0.\boxed{} = 1$

6.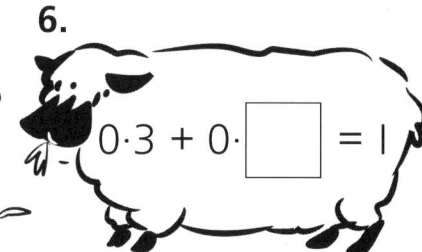
$0.3 + 0.\boxed{} = 1$

7.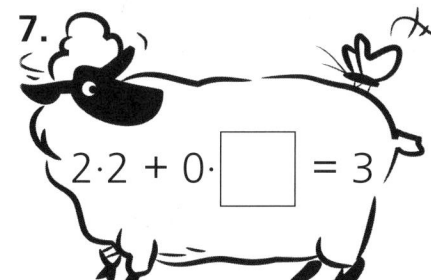
$2.2 + 0.\boxed{} = 3$

8.
$7.6 + 0.\boxed{} = 8$

9.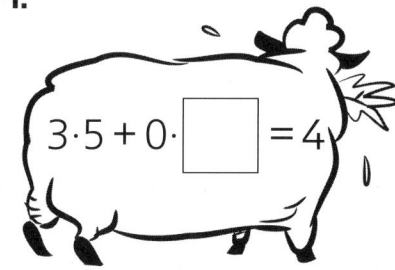
$3.5 + 0.\boxed{} = 4$

10.
$8.4 + 0.\boxed{} = 9$

11.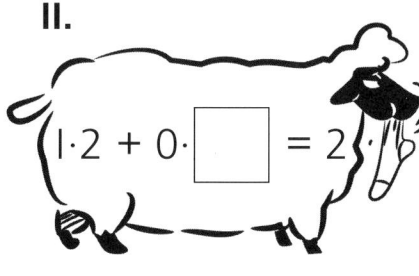
$1.2 + 0.\boxed{} = 2$

12.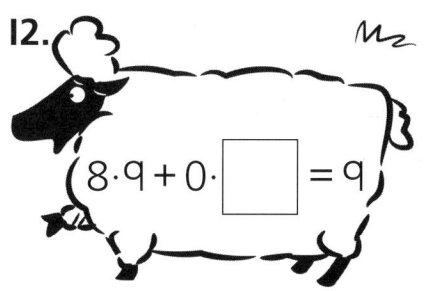
$8.9 + 0.\boxed{} = 9$

13.
$9.4 + 0.\boxed{} = 10$

14.
$2.3 + 0.\boxed{} = 3$

15.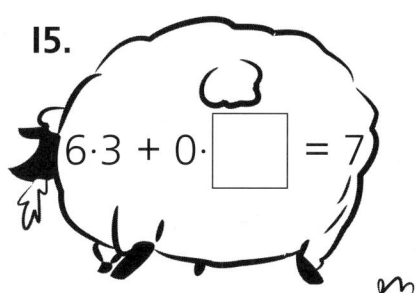
$6.3 + 0.\boxed{} = 7$

43

Abacus Ginn and Company 2001. Copying permitted for purchasing school only. This material is not copyright free.

Name _____

Next whole number

> Write how many to the next whole number.

	0·4 3·6 → 4	4·2 → 5
4·9 → 5	6·2 → 7	7·3 → 8
5·1 → 6	7·5 → 8	9·2 → 10
3·4 → 4	2·2 → 3	4·7 → 5
6·8 → 7	5·6 → 6	8·1 → 9

> Find the coins to make the next whole pound.

£2·48   £3·25

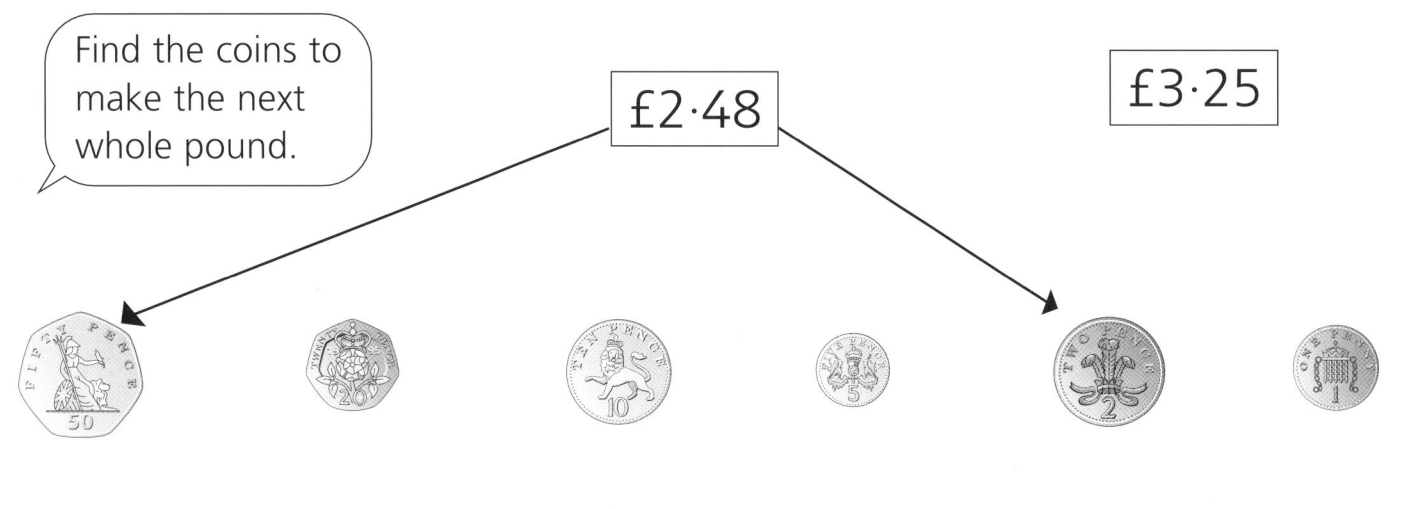

£4·30   £1·85   £3·39

Abacus Ginn and Company 2001. Copying permitted for purchasing school only. This material is not copyright free.

Adding 1-digit and 2-digit numbers

Join pairs
that make 10.

9

3

2

4

7

5

1

5

8

6

Complete
these additions.

$3 + 2 + 8 + 7 =$ 20

Look for
tens.

$3 + 8 + 7 + 8 =$ ☐	$1 + 9 + 1 + 9 =$ ☐	$2 + 6 + 4 + 7 =$ ☐
$1 + 5 + 3 + 5 =$ ☐	$10 + 8 + 6 + 2 =$ ☐	$4 + 12 + 6 =$ ☐
$7 + 19 + 3 =$ ☐	$2 + 5 + 11 + 5 =$ ☐	$1 + 27 + 9 =$ ☐
$4 + 3 + 18 + 7 =$ ☐	$4 + 28 + 6 =$ ☐	$1 + 8 + 14 + 2 =$ ☐
$10 + 6 + 12 + 4 =$ ☐	$2 + 18 + 5 + 5 =$ ☐	$27 + 1 + 9 =$ ☐

Abacus Ginn and Company 2001. Copying permitted for purchasing school only. This material is not copyright free.

Adding 1-digit and 2-digit numbers

Complete these additions.

$4 + 2 + 14 =$ | 20 |

Start with the largest number.

$2 + 3 + 19 =$ ☐	$5 + 4 + 27 =$ ☐	$2 + 29 + 3 =$ ☐
$3 + 18 + 4 =$ ☐	$3 + 28 + 4 =$ ☐	$4 + 17 + 12 =$ ☐
$10 + 18 + 3 =$ ☐	$3 + 16 + 12 =$ ☐	$1 + 5 + 19 + 12 =$ ☐
$2 + 4 + 18 + 13 =$ ☐	$6 + 18 + 8 =$ ☐	$4 + 3 + 16 + 5 =$ ☐
$9 + 7 + 21 + 5 =$ ☐	$6 + 22 + 12 =$ ☐	$1 + 7 + 19 + 12 =$ ☐

Choose three numbers and add them.

How many different totals can you make?

24   18   4

16   3   19   2

Abacus Ginn and Company 2001. Copying permitted for purchasing school only. This material is not copyright free.

Name ————————————————

Adding 1-digit and 2-digit numbers

Complete
these additions.

$2 + 9 + 17 + 8 + 4 =$ ⬚ 40

$3 + 5 + 7 + 19 + 9 =$ ⬚ $2 + 5 + 18 + 5 + 9 =$ ⬚

$6 + 16 + 4 + 9 =$ ⬚ $1 + 28 + 2 + 9 + 8 =$ ⬚

$15 + 5 + 9 + 6 + 1 =$ ⬚ $24 + 9 + 6 + 3 + 4 =$ ⬚

$7 + 18 + 3 + 9 + 2 =$ ⬚ $8 + 9 + 19 + 2 + 6 =$ ⬚

$2 + 6 + 16 + 4 + 9 =$ ⬚ $9 + 17 + 5 + 1 + 9 =$ ⬚

$3 + 9 + 19 + 7 =$ ⬚ $4 + 26 + 6 + 9 + 2 =$ ⬚

How many different
totals can you make
using these cards?

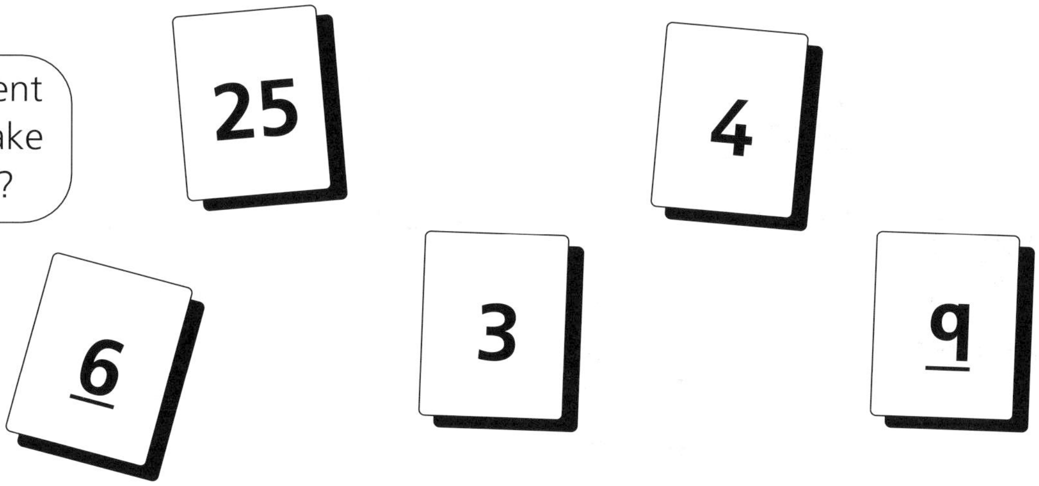

25

4

<u>6</u>

3

<u>9</u>

Abacus Ginn and Company 2001. Copying permitted for purchasing school only. This material is not copyright free.

Name ——————————————

Adding 2-digit numbers

Find the total cost of the stamps.

15p + 15p + 19p + 17p = 66p		

Join numbers which make 100.

 35

 40

25

5

 95

65

60

 20

 80

75

Abacus Ginn and Company 2001. Copying permitted for purchasing school only. This material is not copyright free.

Name ────────────────

Twos, fives and tens

Write the
missing numbers.

| 5 | 10 | 15 | | | | | | | |

| 2 | 4 | | | | | | | | |

| 10 | 20 | | | | | | | | |

Complete these
multiplications.

	8 × 5 = ☐	4 × 2 = ☐
6 × 10 = ☐	3 × 5 = ☐	8 × 2 = ☐
8 × 10 = ☐	6 × 2 = ☐	7 × 5 = ☐
3 × 10 = ☐	7 × 2 = ☐	7 × 10 = ☐
5 × 5 = ☐	4 × 10 = ☐	4 × 5 = ☐

49

Abacus Ginn and Company 2001. Copying permitted for purchasing school only. This material is not copyright free.

Name ——————————————————

Multiplying

Complete this multiplication table.

Double the threes to find the sixes.

	7	2	9	3	8	4	5	10	6
×5	35								
×3					24				
×6					48				

Choose two cards and multiply them.

☐ × ☐ = ☐   ☐ × ☐ = ☐

☐ × ☐ = ☐   ☐ × ☐ = ☐   ☐ × ☐ = ☐

☐ × ☐ = ☐   ☐ × ☐ = ☐   ☐ × ☐ = ☐

☐ × ☐ = ☐   ☐ × ☐ = ☐   ☐ × ☐ = ☐

Abacus Ginn and Company 2001. Copying permitted for purchasing school only. This material is not copyright free.

Square numbers

Write how many squares in each grid.

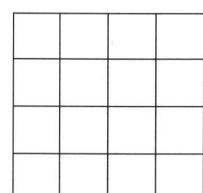 $\boxed{4} \times \boxed{4} = \boxed{16}$

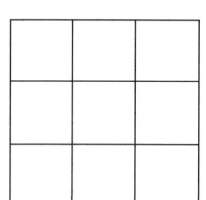 $\boxed{} \times \boxed{} = \boxed{}$

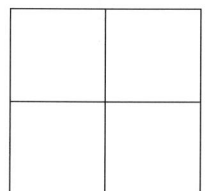 $\boxed{} \times \boxed{} = \boxed{}$

 $\boxed{} \times \boxed{} = \boxed{}$

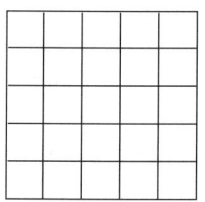 $\boxed{} \times \boxed{} = \boxed{}$

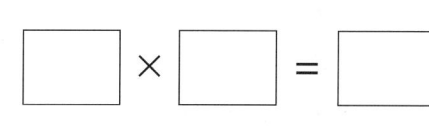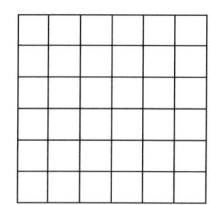 $\boxed{} \times \boxed{} = \boxed{}$

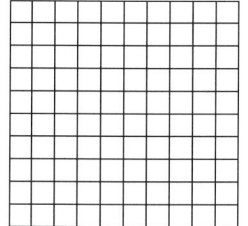 $\boxed{} \times \boxed{} = \boxed{}$

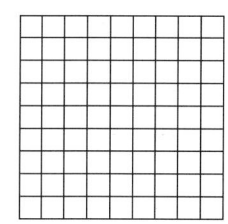 $\boxed{} \times \boxed{} = \boxed{}$

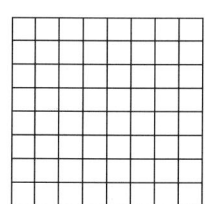 $\boxed{} \times \boxed{} = \boxed{}$

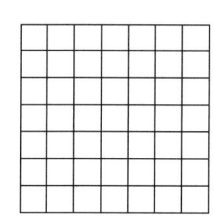 $\boxed{} \times \boxed{} = \boxed{}$

Abacus Ginn and Company 2001. Copying permitted for purchasing school only. This material is not copyright free.

Multiplying

Write the twos, fives and tens in this grid.

Write the square numbers – these go diagonally.

Use your fingers to help you write the nines.

Double the twos to help you write the fours.

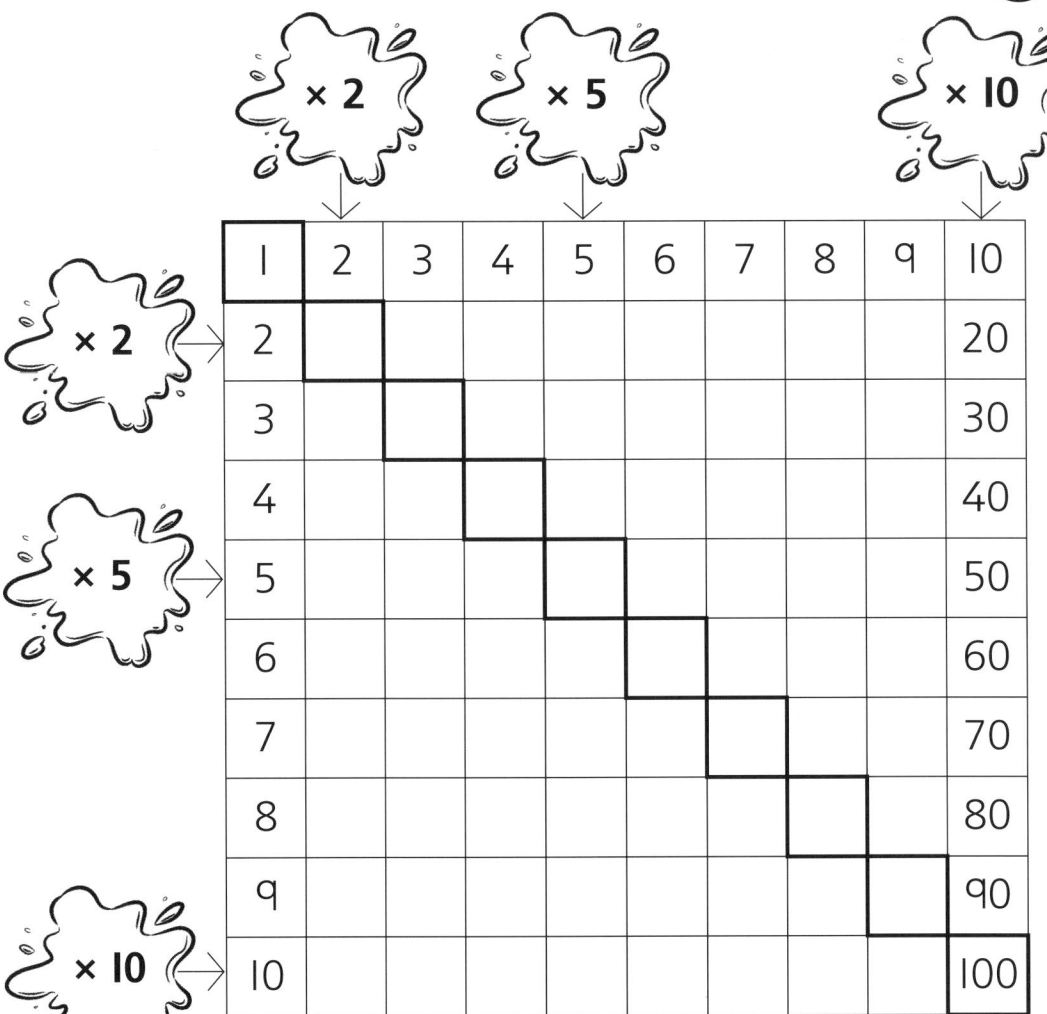

× 2										× 5					× 10
1	2	3	4	5	6	7	8	9	10
2									20
3									30
4									40
5									50
6									60
7									70
8									80
9									90
10									100

Complete this table for the threes.

	1	2	3	4	5	6	7	8	9	10
×3	3	6			15					30

Write these in the multiplication grid above.

Double each answer and write the sixes in the table.

Write any remaining facts.

Abacus Ginn and Company 2001. Copying permitted for purchasing school only. This material is not copyright free.

Name

Multiplying decimals

Complete the
multiplications.

3×42

$3 \times 40 = \boxed{120}$

$3 \times 2 = \boxed{6}$

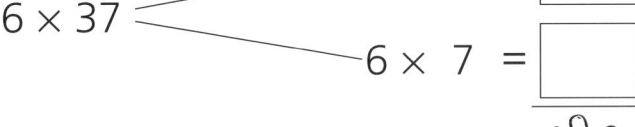 126

5×24

$5 \times 20 = \boxed{}$

$5 \times 4 = \boxed{}$

6×37

$6 \times 30 = \boxed{}$

$6 \times 7 = \boxed{}$

6×17

$6 \times 10 = \boxed{}$

$6 \times 7 = \boxed{}$

4×27

$4 \times 20 = \boxed{}$

$4 \times 7 = \boxed{}$

Complete the decimal
multiplications.

$3 \times 4{\cdot}2$

$3 \times 4{\cdot}0 = \boxed{12{\cdot}0}$

$3 \times 0{\cdot}2 = \boxed{0{\cdot}6}$

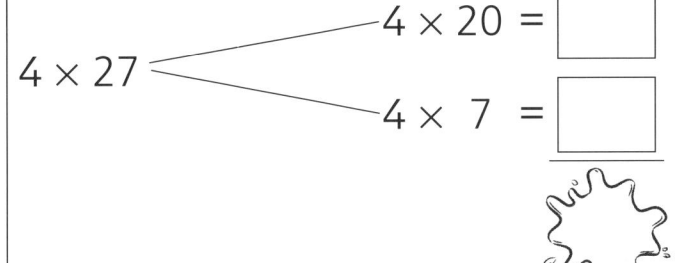 12·6

$4 \times 7{\cdot}8$

$4 \times 7{\cdot}0 = \boxed{}$

$4 \times 0{\cdot}8 = \boxed{}$

$5 \times 6{\cdot}3$

$5 \times 6{\cdot}0 = \boxed{}$

$5 \times 0{\cdot}3 = \boxed{}$

$9 \times 4{\cdot}7$

$9 \times 4{\cdot}0 = \boxed{}$

$9 \times 0{\cdot}7 = \boxed{}$

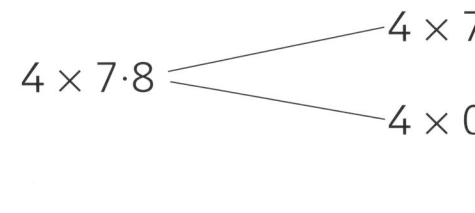

$7 \times 3{\cdot}1$

$7 \times 3{\cdot}0 = \boxed{}$

$7 \times 0{\cdot}1 = \boxed{}$

Abacus Ginn and Company 2001. Copying permitted for purchasing school only. This material is not copyright free.

Name _____

Multiplying decimals

Join the multiplications to the matching answers.

5 × 0·3

7 × 0·8

12 × 0·2

5 × 0·6

1·5

3

5·6

2·4

4

3·6

2·8

4·2

6 × 0·7

8 × 0·5

4 × 0·9

7 × 0·4

Complete the multiplications.

 7 books

£2·40

7 × £2·40 = £ _____

 6 books

£3·40

6 × £3·40 = £ _____

 4 books

£4·80

4 × £4·80 = £ _____

 3 books

£5·60

3 × £5·60 = £ _____

 8 books

£7·10

8 × £7·10 = £ _____

 9 books

£6·30

9 × £6·30 = £ _____

Abacus Ginn and Company 2001. Copying permitted for purchasing school only. This material is not copyright free.

Multiplying decimals

Each bag is 6 times as heavy when full.

Write the weights of the full bags.

empty
2·8 kg

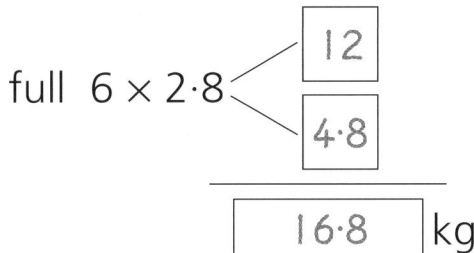

full 6 × 2·8 — 12 / 4·8

16·8 kg

empty
4·3 kg

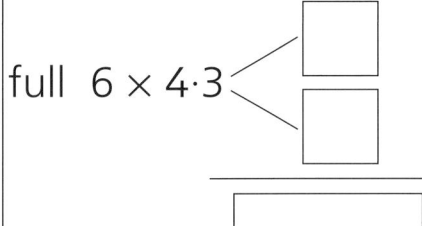

full 6 × 4·3 — ☐ / ☐

☐ kg

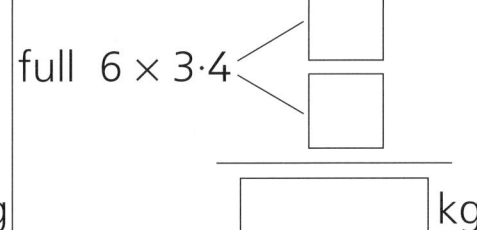

empty
3·4 kg

full 6 × 3·4 — ☐ / ☐

☐ kg

empty
5·1 kg

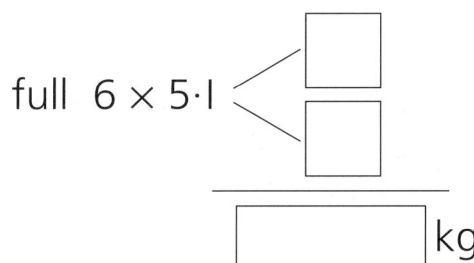

full 6 × 5·1 — ☐ / ☐

☐ kg

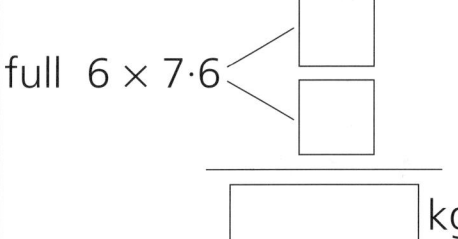

empty
7·6 kg

full 6 × 7·6 — ☐ / ☐

☐ kg

empty
8·2 kg

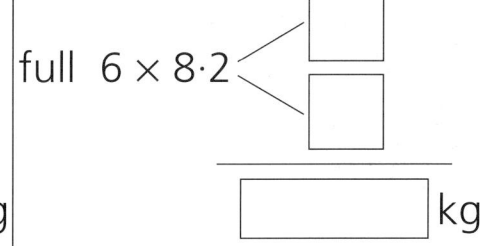

full 6 × 8·2 — ☐ / ☐

☐ kg

Use these cards.

Make these products.

| 0·9 | 5 | 6 | 0·7 | 2 |

a. the largest product

 × ☁ = ☁

b. the smallest product

☁ × ☁ = ☁

c. the product closest to 3

 × ☁ = ☁

Abacus Ginn and Company 2001. Copying permitted for purchasing school only. This material is not copyright free.

Multiplying decimals

Complete these.

4 × 4 · 6 =
16 2·4

○ ○

4 × 2 · 8 =
○ ○

○ ○

4 × 7 · 2 =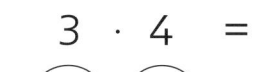
○ ○

○ ○

4 × 3 · 4 =
○ ○

○ ○

4 × 9 · 3 =
○ ○

○ ○

4 × 7 · 9 =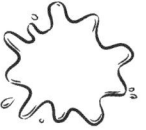
○ ○

○ ○

Choose a balloon number.

Throw the dice.

Multiply by the number thrown.
If you throw a 1, throw again.

a dice

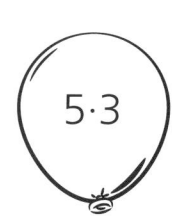
5·3

8·4

7·2

3·6

2·5

4·7

Abacus Ginn and Company 2001. Copying permitted for purchasing school only. This material is not copyright free.

Finding fractions of amounts

Draw the olives
on the slices of pizza.

 12 olives

$\frac{1}{4}$ of 12 = ☐ 3 ☐ per slice

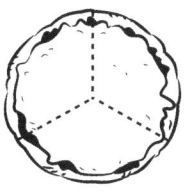 15 olives

$\frac{1}{3}$ of 15 = ☐ per slice

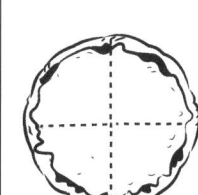 20 olives

$\frac{1}{4}$ of 20 = ☐ per slice

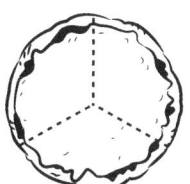 18 olives

$\frac{1}{3}$ of 18 = ☐ per slice

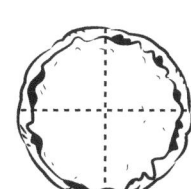 16 olives

$\frac{1}{4}$ of 16 = ☐ per slice

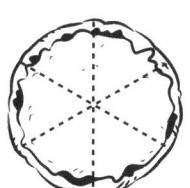 24 olives

$\frac{1}{6}$ of 24 = ☐ per slice

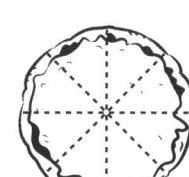 40 olives

$\frac{1}{8}$ of 40 = ☐ per slice

Find $\frac{1}{6}$ of each number of sweets.

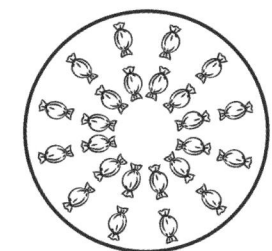

$\frac{1}{6}$ of 24 = ☐

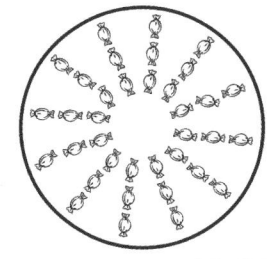

$\frac{1}{6}$ of 36 = ☐

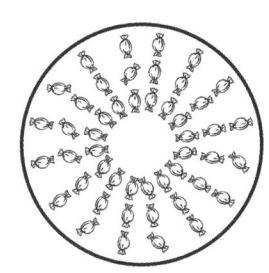

$\frac{1}{6}$ of 48 = ☐

Abacus Ginn and Company 2001. Copying permitted for purchasing school only. This material is not copyright free.

Finding fractions of amounts

Find $\frac{2}{5}$ of each set of biscuits.

15 biscuits

$\frac{1}{5}$ of 15 = ☐

$\frac{2}{5}$ of 15 = ☐

10 biscuits

$\frac{1}{5}$ of 10 = ☐

$\frac{2}{5}$ of 10 = ☐

50 biscuits

$\frac{1}{5}$ of 50 = ☐

$\frac{2}{5}$ of 50 = ☐

30 biscuits

$\frac{1}{5}$ of 30 = ☐

$\frac{2}{5}$ of 30 = ☐

20 biscuits

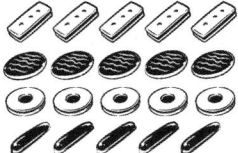

$\frac{1}{5}$ of 20 = ☐

$\frac{2}{5}$ of 20 = ☐

Complete these.

$\frac{1}{6}$ of 30 = ☐

$\frac{5}{6}$ of 30 = ☐

$\frac{1}{3}$ of 18 = ☐

$\frac{2}{3}$ of 18 = ☐

$\frac{1}{4}$ of 16 = ☐

$\frac{3}{4}$ of 16 = ☐

$\frac{1}{7}$ of 28 = ☐

$\frac{2}{7}$ of 28 = ☐

$\frac{1}{8}$ of 48 = ☐

$\frac{3}{8}$ of 48 = ☐

$\frac{1}{3}$ of 24 = ☐

$\frac{2}{3}$ of 24 = ☐

$\frac{1}{9}$ of 72 = ☐

$\frac{4}{9}$ of 72 = ☐

$\frac{1}{10}$ of 190 = ☐

$\frac{3}{10}$ of 190 = ☐

$\frac{1}{20}$ of 100 = ☐

$\frac{9}{20}$ of 100 = ☐

Abacus Ginn and Company 2001. Copying permitted for purchasing school only. This material is not copyright free.

Mixed numbers

Write the missing numbers.

1 cookie →
10 chocolate chips

1½ cookies →

[15] chocolate chips

1 cookie →
12 chocolate chips

1½ cookies →

[   ] chocolate chips

1 cookie →
20 chocolate chips

2¼ cookies →

[   ] chocolate chips

1 cookie →
40 chocolate chips

2⅕ cookies →

[   ] chocolate chips

1 cookie →
35 chocolate chips

3⅐ cookies →

[   ] chocolate chips

Write how many chocolates in each set.

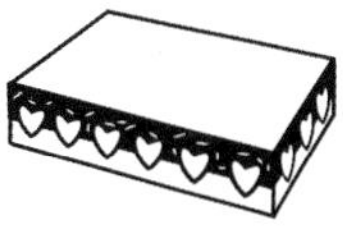

1½ boxes

20 chocolates

[30] chocolates

1¼ boxes

40 chocolates

[   ] chocolates

1⅙ boxes

30 chocolates

[   ] chocolates

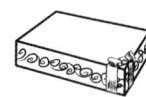

2⅓ boxes

18 chocolates

[   ] chocolates

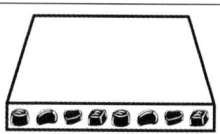

2⅛ boxes

48 chocolates

[   ] chocolates

Abacus Ginn and Company 2001. Copying permitted for purchasing school only. This material is not copyright free.

Name —————————————————

Mixed numbers

Complete these.

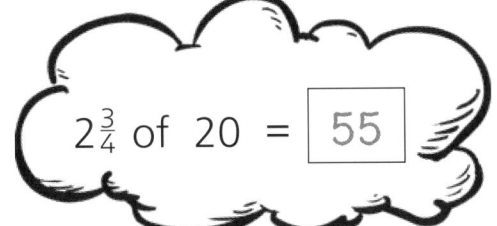

$2\frac{3}{4}$ of 20 = 55

2	of	20	=	40
$\frac{1}{4}$	of	20	=	5
$\frac{3}{4}$	of	20	=	15

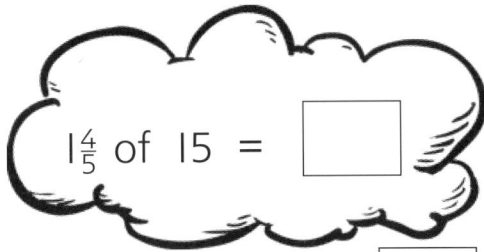

$1\frac{4}{5}$ of 15 = ☐

1	of	15	=	☐
$\frac{1}{5}$	of	15	=	☐
$\frac{4}{5}$	of	15	=	☐

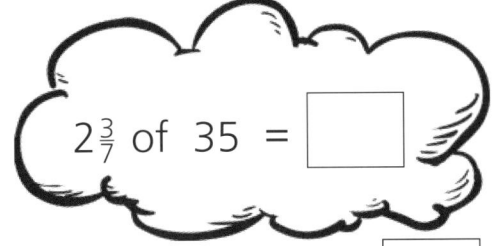

$2\frac{3}{7}$ of 35 = ☐

2	of	35	=	☐
$\frac{1}{7}$	of	35	=	☐
$\frac{3}{7}$	of	35	=	☐

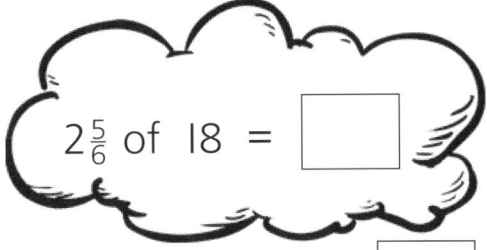

$2\frac{5}{6}$ of 18 = ☐

2	of	18	=	☐
$\frac{1}{6}$	of	18	=	☐
$\frac{5}{6}$	of	18	=	☐

Complete the stars by writing the amounts at each point.

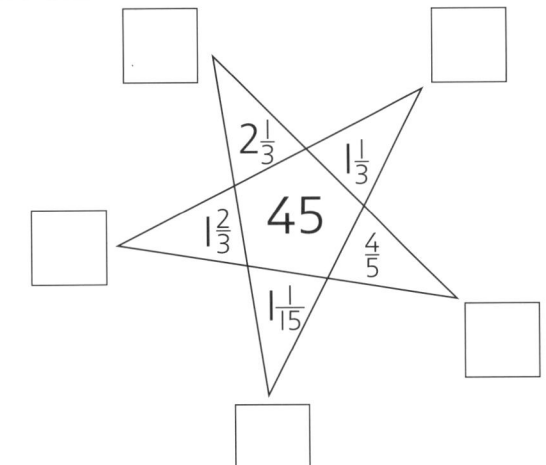

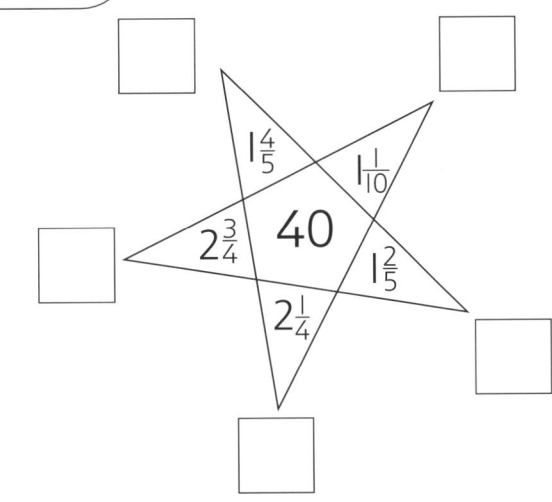

Abacus Ginn and Company 2001. Copying permitted for purchasing school only. This material is not copyright free.

Finding tenths

Complete the divisions.

270 ÷ 10 = ☐ 27

27 ÷ 10 = ☐ 2·7

350 ÷ 10 = ☐

35 ÷ 10 = ☐

420 ÷ 10 = ☐

42 ÷ 10 = ☐

150 ÷ 10 = ☐

15 ÷ 10 = ☐

680 ÷ 10 = ☐

68 ÷ 10 = ☐

490 ÷ 10 = ☐

49 ÷ 10 = ☐

110 ÷ 10 = ☐

11 ÷ 10 = ☐

770 ÷ 10 = ☐

77 ÷ 10 = ☐

190 ÷ 10 = ☐

19 ÷ 10 = ☐

230 ÷ 10 = ☐

23 ÷ 10 = ☐

Colour $\frac{1}{10}$ of each grid.

a.
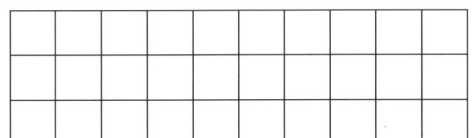

b.

c.
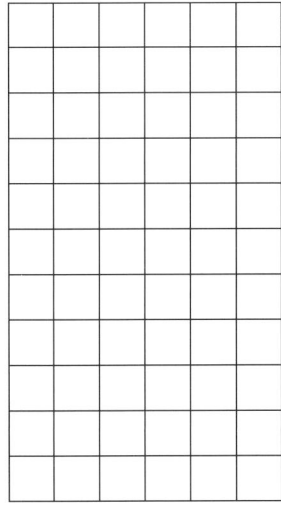

d.
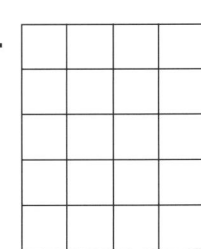

e.

Abacus Ginn and Company 2001. Copying permitted for purchasing school only. This material is not copyright free.

Finding 10% of numbers

Colour 10% of each set of cubes.

10% of 110 =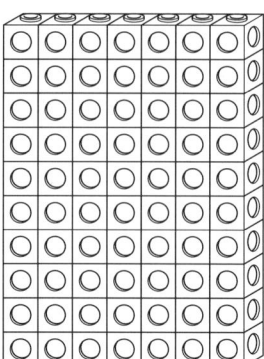

10% of 70 =

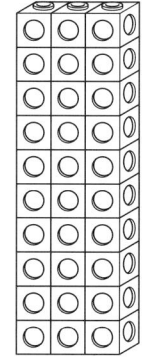

10% of 50 =

10% of 120 =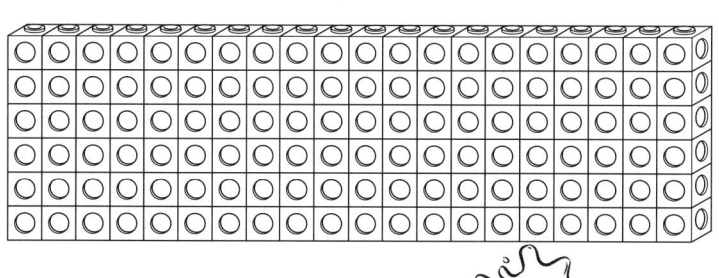

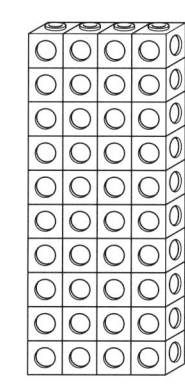

10% of 40 =

10% of 30 =

Write 10% of each number.

50

5

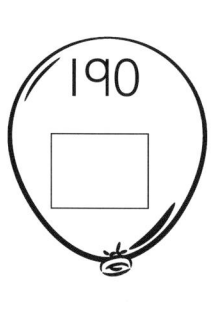

190

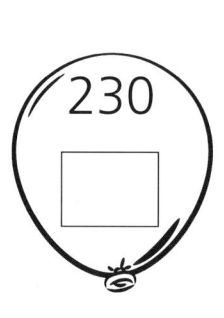

230

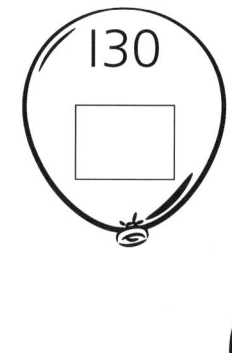

130

70

360

440

240

Abacus Ginn and Company 2001. Copying permitted for purchasing school only. This material is not copyright free.

Finding 10% of amounts

Find 10% of each amount.

Reduce each price by 10%.

Write each new price.

10% of £40 = £ 4

£40 – £ 4 = £ 36

£40

10% of £80 = £ ☐

£80 – £ ☐ = £ ☐

£80

10% of £75 = £ ☐

£75 – £ ☐ = £ ☐

£75

10% of £30 = £ ☐

£30 – £ ☐ = £ ☐

£30

10% of £45 = £ ☐

£45 – £ ☐ = £ ☐

£45

10% of £60 = £ ☐

£60 – £ ☐ = £ ☐

£60

10% of £25 = £ ☐

£25 – £ ☐ = £ ☐

£25

10% of £250 = £ ☐

£250 – £ ☐ = £ ☐

£250

10% of £320 = £ ☐

£320 – £ ☐ = £ ☐

£320

Abacus Ginn and Company 2001. Copying permitted for purchasing school only. This material is not copyright free.

Finding percentages

Complete these.

10% of £33 = £ 3·30

30% of £33 = £ 9·90

10% of £45 = £ ☐	10% of £25 = £ ☐
20% of £45 = £ ☐	90% of £25 = £ ☐
10% of £62 = £ ☐	10% of £16 = £ ☐
80% of £62 = £ ☐	70% of £16 = £ ☐
10% of £52 = £ ☐	10% of £15 = £ ☐
50% of £52 = £ ☐	40% of £15 = £ ☐

A 30% deposit is needed for each holiday.

Write how much each person pays.

£450

10% of £450 = £ ☐

30% of £450 = £ ☐

£210

10% of £210 = £ ☐

30% of £210 = £ ☐

£390

10% of £390 = £ ☐

30% of £390 = £ ☐

£1260

10% of £1260 = £ ☐

30% of £1260 = £ ☐

Abacus Ginn and Company 2001. Copying permitted for purchasing school only. This material is not copyright free.